Introduction to Mineralogy

Introduction to Mineralogy

Cyaden Ross

www.callistoreference.com

Callisto Reference,
118-35 Queens Blvd., Suite 400,
Forest Hills, NY 11375, USA

Visit us on the World Wide Web at:
www.callistoreference.com

ISBN: 978-1-64116-237-1 (Hardback)

Printed in China.

Cataloging-in-Publication Data

Introduction to Mineralogy / Cyaden Ross.
p. cm.
Includes bibliographical references and index.
ISBN 978-1-64116-237-1
1. Mineralogy. 2. Geology. I. Ross, Cyaden.
QE363.2 .I58 2019
549--dc23

Table of Contents

Preface

Minerals are formed through various processes such as sublimation from volcanic gases, crystallization from an igneous magma or lava, recrystallization due to metasomatism and metamorphic processes, besides many others. Minerals can be classified into different groups such as carbonates, sulfides, phosphates, silicates, halides, chromates, etc. Mineralogy is a field of geology. It is concerned with the scientific study of the crystal structure, chemistry and physical properties of minerals. The classification of minerals, processes of origin, formation, geographical distribution and utilization are also within the scope of this field. The properties of minerals such as density, solubility, radioactivity, mechanical cohesion, etc. are studied in detail. This book unfolds the innovative aspects of mineralogy, which will be crucial for the holistic understanding of the subject matter. The topics included in this book are of the utmost significance and bound to provide incredible insights to the readers. It will serve as a reference to professionals and students engaged in this field.

To facilitate a deeper understanding of the contents of this book a short introduction of every chapter is written below:

Chapter 1- Mineralogy is the study of the physical properties, crystal structure and chemistry of minerals and mineralized artifacts. The topics covered in this chapter discuss the fundamentals of mineralogy. It includes topics like mineral, ore mineral, Mohs scale of mineral hardness, streak, type specimen, etc.

Chapter 2- Industrial minerals are mined for their use in industries or other commercial activities depending on their physical or chemical properties. The different types of industrial minerals such as pumice, kaolinite, mica, kyanite, dolomite, bentonite, feldspar, etc. have been discussed in great detail in this chapter.

Chapter 3- The study of rocks and minerals by an analysis of their optical properties is known as optical mineralogy. Some of the properties measured are refractive index, birefringence, extinction angle, etc. This chapter provides an overview of optical mineralogy and examines the fundamentals of petrographic microscope and thin section.

Chapter 4- Magnetic mineralogy is the science concerned with the study of the magnetic properties of minerals. How a mineral contributes to the total magnetism of a rock is dependent on the kind of magnetic order or disorder. This chapter closely examines the key concepts of magnetic mineralogy with topics such as magnetite, greigite, hematite, kamacite, etc. for an extensive understanding of the subject.

Chapter 5- Gemology is a branch of mineralogy, which is concerned with artificial and natural gemstones. This chapter has been carefully written to provide an easy understanding of gemology and the study of gemstone minerals. Many topics related to this field such as cardinal gem, asterism and chatoyancy have been extensively explored in this chapter.

Chapter 6- Crystallography is the science concerned with the determination of the arrangement of atoms in a crystalline solid. In order to study the role of crystallography, it is necessary to understand the fundamentals of lattice constant, crystal systems and their types, crystal family and crystal habit, besides others which have been extensively dealt in this chapter.

I owe the completion of this book to the never-ending support of my family, who supported me throughout the project.

Cyaden Ross

Chapter 1

Understanding Mineralogy

Mineralogy is the study of the physical properties, crystal structure and chemistry of minerals and mineralized artifacts. The topics covered in this chapter discuss the fundamentals of mineralogy. It includes topics like mineral, ore mineral, Mohs scale of mineral hardness, streak, type specimen, etc.

Mineralogy is a scientific discipline that is concerned with all aspects of minerals, including their physical properties, chemical composition, internal crystal structure, and occurrence and distribution in nature and their origins in terms of the physicochemical conditions of formation.

Like many sciences, mineralogy has its origins in several ancient civilizations, and it has been concerned primarily with the various methods of classification of minerals for most of its history. Modern-day mineralogy has been expanded by advances in other sciences, such as biology and chemistry, to shed even more light on the nature of the materials that form the earth we live on.

The simplest way to study and classify minerals is to do so based on their physical attributes. These can include the microscopic structure of the mineral crystals, the way in which the mineral cleaves when struck, and its hardness. The Mohs scale of mineral hardness describes this particular property.

At the low end of the Mohs scale, the soft graphite used in pencils scores a one out of ten, whereas a diamond, the hardest earthly substance, scores a ten. It is interesting to note that both of these minerals are made of the same element, namely carbon. The differences in hardness and appearance are due only to the difference in the way the individual atoms are organized.

There are over 4,000 different types of minerals known to exist. Most of these are classified as rare or extremely rare, and only about 150 are present in large amounts. Another 50 to 100 are found only occasionally. Minerals not only comprise much of the earth's crust, but many are essential to good health. Many types also provide us with important building materials, and the components for some kinds of machinery.

Mineral

Minerals are solid substances that are present in nature and can be made of one element or more elements combined together (chemical compounds). The word "mineral" is used inconsistently in geology.

In mining, anything obtained from the ground and used by man is considered to be a "mineral commodity" or a "mineral material." These include: crushed stone, which is a manufactured product made from crushed rocks; lime, which is a manufactured product made from limestone or marble (both composed of the mineral calcite); coal which is organic; oil and gas which are organic fluids; rocks such as granite that are mixtures of minerals; and, rocks such as obsidian which do not have a definite composition and ordered internal structure.

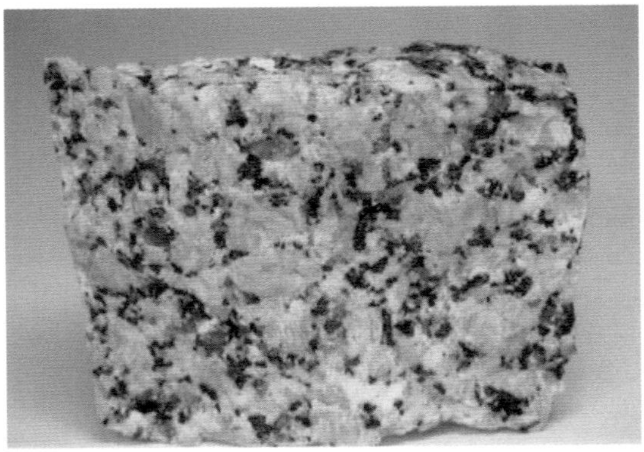

Minerals in rocks: Most rocks are aggregates of minerals. This rock, a granite pegmatite, is a mixture of mineral grains. It contains pink orthoclase, milky quartz, black hornblende and black biotite.

Mineral Commodities in Industry

2010 United States Mineral Commodity Consumption	
Mineral Commodity	Millions of Metric Tons
Crushed Stone	1,200.0
Sand and Gravel	786.1
Salt	55.8
Iron Ore	48.0
Phosphate Rock	30.5
Gypsum	22.5
Clays	21.3
Dimension Stone	14.0
Lime	18.5
Sulfur	11.30
Bauxite	8.4
Potash	5.6
Soda Ash	5.2
Barite	2.66
Copper	1.74
Lead	1.40
Values above are estimates of mineral commodity consumption from the United States Geological Survey. Many other commodities could be added to this table.	

The construction industry is the largest consumer of mineral commodities. Crushed stone is used for foundations, road base, concrete, and drainage. Sand and gravel are used in concrete and foundations. Clays are used to make cement, bricks, and tile. Iron ore is used to make reinforcing rods, steel beams, nails, and wire. Gypsum is used to make drywall. Dimension stone is used for facing, curbing, flooring, stair treads, and other architectural work. These are just a few of the many uses for these commodities in construction.

In agriculture, phosphate rock and potash are used to make fertilizer. Lime is used as an acid-neutralizing soil treatment. Mineral nutrients are added to animal feed.

The chemical industry uses large amounts of salt, lime, and soda ash. Large amounts of metals, clay, and mineral fillers/extenders are used in manufacturing.

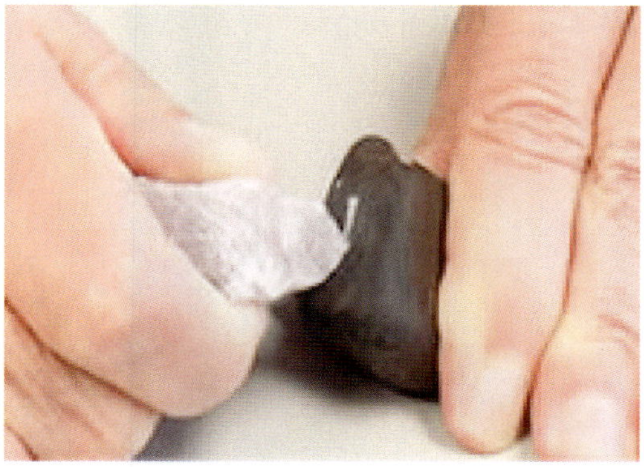

Mohs Hardness Scale is a set of reference minerals used for classroom hardness testing. Determining the hardness of a mineral is one of the most important tests used in mineral identification.

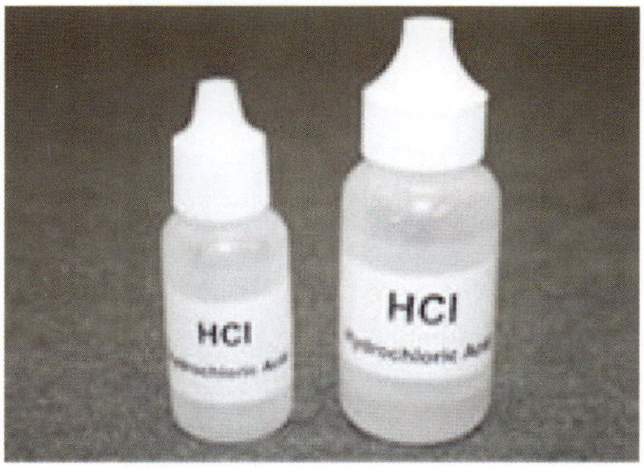

The Acid Test: Geologists use dilute hydrochloric acid to identify carbonate minerals. Carbonate minerals will effervesce with various levels of vigor in contact with the acid.

Physical Properties of Minerals

There are approximately 4000 different minerals, and each of those minerals has a unique set of physical properties. These include: color, streak, hardness, luster, diaphaneity, specific gravity, cleavage, fracture, magnetism, solubility, and many more. These physical properties are useful for identifying minerals. However, they are much more important in determining the potential industrial uses of the mineral. Let's consider a few examples.

The mineral talc, when ground into a powder, is perfectly suited for use as a foot powder. It is a soft, slippery powder so it will not cause abrasion. It has the ability to absorb moisture, oils, and odor. It adheres to the skin and produces an astringent effect - yet it washes off easily. No other mineral has a set of physical properties that are as suitable for this purpose.

The mineral halite, when crushed into small grains, is perfectly suited for flavoring food. It has a salty taste that most people find pleasing. It dissolves quickly and easily, allowing its flavor to spread through the food. It is soft, so if some does not dissolve it will not damage your teeth. No other mineral has physical properties that are better suited for this use.

The mineral gold is perfectly suited for use in jewelry. It can be easily shaped into a custom item of jewelry by a craftsperson. It has a pleasing yellow color that most people enjoy. It has a bright luster that does not tarnish. Its high specific gravity gives it a nice «heft» that is preferred by most people over lighter metals. Other metals can be used to make jewelry, but these properties make gold an overwhelming favorite. (Some people might add that gold›s rarity and value are two additional properties that make it desirable for jewelry. However, rarity is not a property, and its value is determined by supply and demand).

Star sapphire: A deep blue star sapphire 8 mm x 6 mm cabochon from Thailand.

Inclusions of rutile within the stone align with the crystallographic axis of the corundum to produce the star - which is only clearly visible and centered when the back of the stone is cut at 90 degrees to the C-axis of the crystal. This stone has been heat treated to darken the stone and enhance visibility of the star.

Mineral Information

Minerals on Geology.com: Links to content about minerals on the geology.com website.

USGS Mineral Information: Links to information about minerals on the United States Geological Survey website.

State and National Geological Surveys: Links to geological survey websites, most of which contain information about minerals found and produced in their location.

USGS Mineral Resources Program: The sole Federal source of scientific information on mineral potential, production, consumption, and environmental effects.

The Mineral and Locality Database: The largest mineralogy database and reference website on the internet.

Importance of Physical Properties

The primary characteristics of a mineral that determine its physical properties are its composition and the strength of the bonds in its ordered internal structure. Here are some examples:

Galena, a lead sulfide, has a much higher specific gravity than bauxite, an aluminum hydroxide. This difference is because of their composition. Lead is much heavier than aluminum.

Diamond and graphite both consist of pure carbon. Diamond is the hardest natural mineral, and graphite is one of the softest. This difference occurs because of the types of bonds connecting the carbon atoms in their mineral structures. Each carbon atom in diamond is bonded to four other carbon atoms with strong covalent bonds. Graphite has a sheet structure in which atoms within the sheets are bonded to one another with strong covalent bonds, but the bonds between the sheets are weak electrical bonds. When graphite is scratched the weak bonds fail easily, making it a soft mineral.

The gemstones ruby and sapphire are color variations of the mineral corundum. These color differences are caused by composition. When corundum contains trace amounts of chromium, it exhibits the red color of a ruby. However, when it contains trace amounts of iron or titanium, it exhibits the blue color of sapphire. If, at the time of crystallization, enough titanium is present to form tiny crystals of the mineral rutile, a star sapphire may form. This occurs when tiny crystals of rutile align systematically within the crystalline structure of the corundum to give it a silky luster that might produce a "star" that aligns with the primary crystallographic axis.

Index Mineral

Some common minerals in metamorphic rocks are shown in figure, arranged in order of the temperature ranges within which they tend to be stable. The upper and lower limits of the ranges are intentionally vague because these limits depend on a number of different factors, such as the pressure, the amount of water present, and the overall composition of the rock. Even though the limits of the stability ranges are vague, the stability range of each mineral is still small enough that the minerals can be used as markers for those metamorphic conditions. Minerals which make good markers of specific ranges of metamorphic conditions are called index minerals. When geologists are examining the metamorphic rocks in a region, they can use the index minerals to map out zones that experienced different pressures and temperatures.

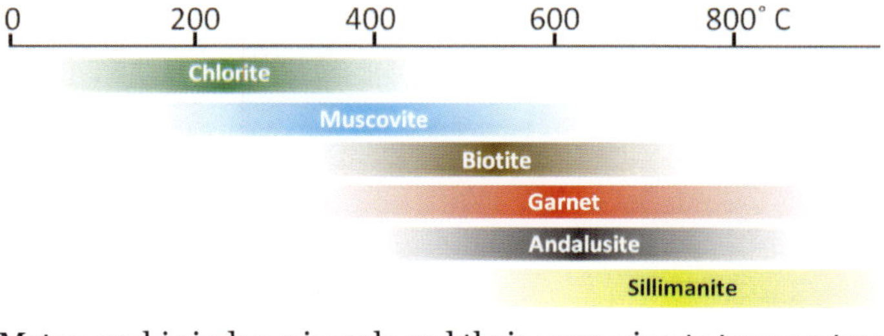

Figure: Metamorphic index minerals and their approximate temperature ranges.

The Meguma Terrane of Nova Scotia: An Example of How to Use Index Minerals.

The southern and southwestern parts of Nova Scotia were regionally metamorphosed during the Devonian Acadian Orogeny (around 400 Ma), when a relatively small continental block (the Meguma Terrane) was pushed up against the existing eastern margin of North America. The clastic sedimentary rocks within this terrane were variably metamorphosed. Figure below is a map of zones where different index minerals can be found. This allows us to see where metamorphism was stronger or weaker.

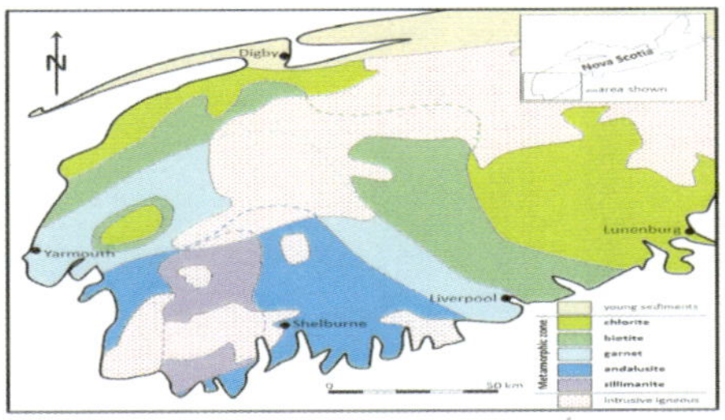

Figure: Regional metamorphic zones in the Meguma Terrane of southwestern Nova Scotia.

The strongest metamorphism (the sillimanite zone) is in the southwest. Progressively weaker metamorphism exists toward the east and north. The rocks of the sillimanite zone were likely heated to over 700° C, and therefore must have buried to depths between 20 km and 25 km. The surrounding lower-grade rocks were not buried as deep, and the rocks within the peripheral chlorite zone were likely not buried to more than about 5 km.

A probable explanation for this pattern is that the area with the highest-grade rocks was buried beneath the central part of a mountain range formed by the collision of the Meguma Terrane with North America. As is the case with all mountain ranges, the crust became thickened as the mountains grew, and it was pushed farther down into the mantle than the surrounding crust. This happens because Earth's crust is floating on the underlying mantle. As the formation of mountains adds weight, the crust in that area sinks farther down into the mantle to compensate for the added weight. The likely pattern of metamorphism in this situation is shown in cross-section in Figure (a) below. The mountains were eventually eroded (over tens of millions of years), allowing the crust to rebound upward and exposing the metamorphic rock.

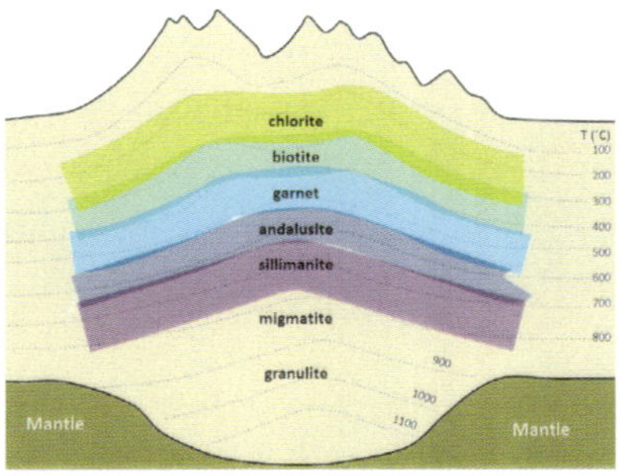

Figure (a): Schematic cross-section through the Meguma Terrane during the Devonian. The crust is thickened underneath the mountain range to compensate for the added weight of the mountains above. Temperature contours are shown, and the metamorphic zones are depicted using colours similar.

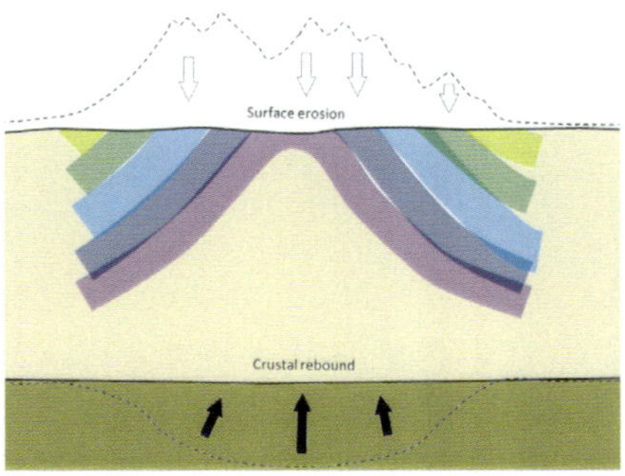

Figure (b): Schematic present-day cross-section through the Meguma Terrane. The mountains have been eroded. As they lost mass the base of the crust gradually rebounded, pushing up the core of the metamorphosed region so that the once deeply buried metamorphic zones are now exposed at surface.

Building a narrative for the metamorphism in Nova Scotia's Meguma Terrane is just one example of how index minerals can be used.

Scottish Metamorphic Zones

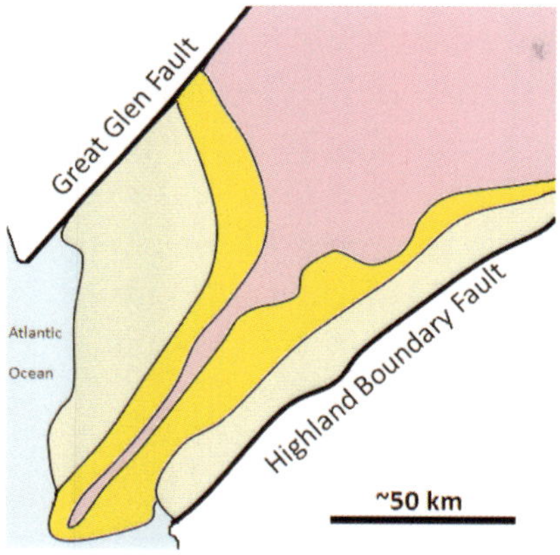

The map shown here represents the part of western Scotland between the Great Glen Fault and the Highland Boundary Fault. The shaded areas are metamorphic rock, and the three metamorphic zones represented are garnet, chlorite, and biotite.

Label the three coloured areas of the map with the appropriate zone names (garnet, chlorite, and biotite).

Indicate which part of the region was likely to have been buried the deepest during metamorphism.

British Geologist George Barrow studied this area in the 1890s and was the first person anywhere to map metamorphic zones based on their mineral assemblages. This pattern of metamorphism is sometimes referred to as "Barrovian."

Examples of Index Minerals

The most widely used index minerals are, in ascending order of pressure/temperature, are biotite, the zeolites, chlorite, prehnite, biotite, hornblende, garnet, glaucophane, staurolite, sillimanite, and glaucophane.

When these minerals are found in particular types of rocks, they can indicate the minimum amount of pressure and temperature the rock has experienced.

For example, slate, when it undergoes metamorphosis, changes first to phyllite, then to schist, and finally to gneiss. When slate is seen to contain chlorite, it is understood to have undergone a low grade of metamorphosis.

Mudrock, a sedimentary rock, contains quarts at all grades of metamorphosis. Other minerals, however, are added as the rock undergoes different "zones" of metamorphosis. The minerals are added in the following order: biotite, garnet, staurolite, kyanite, sillimanite. If a piece of mudrock contains garnet but no kyanite, it has probably undergone only a low grade of metamorphosis. If, however, it contains sillimanite, it has undergone extreme metamorphosis.

Rock

In geology, Rock is a naturally occurring and coherent aggregate of one more mineral. Such aggregates constitute the basic unit of which the solid Earth is comprised and typically form recognizable and mappable volumes. Rocks are commonly divided into three major classes according to the processes that resulted in their formation. These classes are (1) igneous rocks, which have solidified from molten material called magma; (2) sedimentary rocks, those consisting of fragments derived from preexisting rocks or of materials precipitated from solutions; and (3) metamorphic rocks, which have been derived from either igneous or sedimentary rocks under conditions that caused changes in mineralogical composition, texture, and internal structure. These three classes, in turn, are subdivided into numerous groups and types on the basis of various factors, the most important of which are chemical, mineralogical, and textural attributes.

Igneous Rock

Igneous rock, or magmatic rock, is one of the three main rock types.

Igneous rock is formed through the cooling and solidification of magma or lava. The magma can be derived from partial melts of existing rocks in either a planet's mantle or crust. Typically, the melting is caused by one or more of three processes: an increase in temperature, a decrease in pressure, or a change in composition. Solidification into rock occurs either below the surface as intrusive rocks or on the surface as extrusive rocks. Igneous rock may form with crystallization to form granular, crystalline rocks, or without crystallization to form natural glasses.

There are two basic types:

1. Intrusive igneous rocks: crystallize below Earth's surface, and the slow cooling that occurs there allows large crystals to form. Examples of intrusive igneous rocks are diorite, gabbro, granite, pegmatite, and peridotite.

2. Extrusive igneous rocks: erupt onto the surface, where they cool quickly to form small crystals. Some cool so quickly that they form an amorphous glass. These rocks include andesite, basalt, obsidian, pumice, rhyolite, scoria, and tuff.

Sample of basalt (an extrusive igneous rock), found in Massachusetts.

Granite.

Gabbro specimen; Rock Creek Canyon, eastern Sierra Nevada, California.

Metamorphic Rock

Metamorphic rocks arise from the transformation of existing rock types, in a process called metamorphism, which means "change in form".

The original rock (protolith) is subjected to heat and pressure causing profound physical and chemical change. The protolith may be a sedimentary rock, an igneous rock or another older metamorphic rock.

Metamorphic rocks make up a large part of the Earth's crust and form 12% of the Earth's current land surface. They are classified by texture and by chemical and mineral assemblage (metamorphic facies). They may be formed simply by being deep beneath the Earth's surface, subjected to high temperatures and the great pressure of the rock layers above it. They can form from tectonic processes such as continental collisions, which cause horizontal pressure, friction and distortion. They are also formed when rock is heated up by the intrusion of hot molten rock called magma from the Earth's interior.

There are two basic types of metamorphic rocks:

Foliated metamorphic rocks: such as gneiss, phyllite, schist, and slate have a layered or banded appearance that is produced by exposure to heat and directed pressure.

Non-foliated metamorphic rocks: such as hornfels, marble, quartzite, and novaculite do not have a layered or banded appearance.

Quartzite.

Gneiss.

Folded foliation in a metamorphic rock from near Geirangerfjord, Norway.

Sedimentary Rock

Sedimentary rock is one of the three main rock groups (along with igneous and metamorphic rocks) and is formed in four main ways: by the deposition of the weathered remains of other rocks (known as 'clastic' sedimentary rocks); by the accumulation and the consolidation of sediments; by the deposition of the results of biogenic activity; and by precipitation from solution.

Sedimentary rocks include common types such as chalk, limestone, sandstone, clay and shale.

Sedimentary rocks cover 75% of the Earth's surface.

Four basic processes are involved in the formation of a clastic sedimentary rock: weathering (erosion) caused mainly by friction of waves, transportation where the sediment is carried along by a current, deposition and compaction where the sediment is squashed together to form a rock of this kind.

Sedimentary rocks are formed from overburden pressure as particles of sediment are deposited out of air, ice, or water flows carrying the particles in suspension.

As sediment deposition builds up, the overburden (or 'lithostatic') pressure squeezes the sediment into layered solids in a process known as lithification ('rock formation') and the original connate fluids are expelled.

The term diagenesis is used to describe all the chemical, physical, and biological changes, including cementation, undergone by a sediment after its initial deposition and during and after its lithification, exclusive of surface weathering.

There are three basic types of sedimentary rocks

1. Clastic sedimentary rocks: such as breccia, conglomerate, sandstone, siltstone, and shale are formed from mechanical weathering debris.
2. Chemical sedimentary rocks: such as rock salt, iron ore, chert, flint, some dolomites, and some limestones, form when dissolved materials precipitate from solution.
3. Organic sedimentary rocks: such as coal, some dolomites, and some limestones, form from the accumulation of plant or animal debris.

Sandstone.

Calcitic Limestone.

Chert

Lump of Coal

Breccia

Rock Cycle

Geologic materials—mineral crystals and their host rock types—are cycled through various forms. The process depends on temperature, pressure, time, and changes in environmental conditions in the Earth's crust and at its surface. The rock cycle illustrated in Figure below reflects the basic relationships among igneous, metamorphic, and sedimentary rocks. Erosion includes weathering (the physical and chemical breakdown of minerals) and transportation to a site of deposition. Diagenesis is, as previously explained, the process of forming sedimentary rock by compaction and natural cementation of grains, or crystallization from water or solutions, or recrystallization. The conversion of sediment to rock is termed lithification.

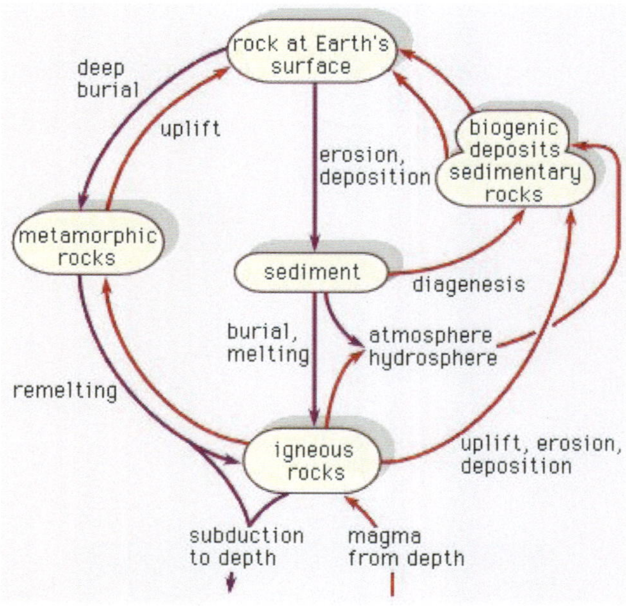

Figure: The rock cycle

Texture

The texture of a rock is the size, shape, and arrangement of the grains (for sedimentary rocks) or crystals (for igneous and metamorphic rocks). Also of importance are the rock's extent of homogeneity (*i.e.*, uniformity of composition throughout) and the degree of isotropy. The latter is the extent to which the bulk structure and composition are the same in all directions in the rock.

obsidian · porphyry · calico, or laminated sandstone · coquina, or shell limestone

breccia · banded gneiss · talc schist · serpentine

Rocks have many different textures. Layered sandstone produces a gritty texture, whereas coquina may be rough with cemented shells occasionally producing a sharp edge. Likewise, breccia, which contains pieces of other rocks that have been cemented together, and porphyry, which contains interlocking mineral crystals, tend to be rough. In contrast, obsidian tends to have a smooth glassy feel, whereas serpentine may feel platy or fibrous, and talc schist often feels greasy. On the other hand, the texture of gneiss is often described by its distinct banding.

Analysis of texture can yield information about the rock's source material, conditions and environment of deposition (for sedimentary rock) or crystallization and recrystallization (for igneous and metamorphic rock, respectively), and subsequent geologic history and change.

Classification by Grain or Crystal Size

The common textural terms used for rock types with respect to the size of the grains or crystals, are given in the Table. The particle-size categories are derived from the Udden-Wentworth scale developed for sediment. For igneous and metamorphic rocks, the terms are generally used as modifiers—*e.g.*, medium-grained granite. Aphanitic is a descriptive term for small crystals, and phaneritic for larger ones. Very coarse crystals (those larger than 3 centimetres, or 1.2 inches) are termed pegmatitic.

Table 32: Common Textural Terms for Rocks*

size (in millimetres)	igneous and metamorphic		sedimentary: sediment	sedimentary: rock
256	phaneritic	very coarse (pegmatitic)	boulder	conglomerate
128			cobble	
64				
32			pebble	
16		coarse		
8				
4		medium		
2			granule	
1			sand	coarse sandstone
1/2		fine		
1/4				fine sandstone
1/8				
1/16	aphanitic	dense		
1/32			silt	siltstone
1/64				
1/128				
1/256			clay	shale

size (in millimetres)	pyroclastic: sediment	pyroclastic: rock
256–32	block bomb	breccia
16–2	lapilli, cinder	lapilli
1–1/8	coarse ash	coarse tuff
1/16–1/256	fine ash	fine tuff

*Diagonal lines in the table reflect the variability in size limits of certain grades resulting from the use of different values by different authors.

For sedimentary rocks, the broad categories of sediment size are coarse (greater than 2 millimetres, or 0.08 inch), medium (between 2 and $^1/_{16}$millimetres), and fine (under $^1/_{16}$ millimetre). The latter includes silt and clay, which both have a size indistinguishable by the human eye and are also termed dust. Most shales (the lithified version of clay) contain some silt. Pyroclastic rocks are those formed from clastic (from the Greek word for broken) material ejected from volcanoes. Blocks are fragments broken from solid rock, while bombs are molten when ejected.

Porosity

The term rock refers to the bulk volume of the material, including the grains or crystals as well as the contained void space. The volumetric portion of bulk rock that is not occupied by grains, crystals, or natural cementing material is termed porosity. That is to say, porosity is the ratio of void volume to the bulk volume (grains plus void space). This void space consists of pore space between grains or crystals, in addition to crack space. In sedimentary rocks, the amount of pore space depends on the degree of compaction of the sediment (with compaction generally increasing with depth of burial), on the packing arrangement and shape of grains, on the amount of cementation, and on the degree of sorting. Typical cements are siliceous, calcareous or carbonate, or iron-bearing minerals.

Sorting is the tendency of sedimentary rocks to have grains that are similarly sized—*i.e.,* to have a narrow range of sizes. Poorly sorted sediment displays a wide range of grain sizes and hence has decreased porosity. Well-sorted indicates a grain size distribution that is fairly uniform. Depending on the type of close-packing of the grains, porosity can be substantial. It should be noted that in engineering usage—*e.g.,* geotechnical or civil engineering—the terminology is phrased oppositely and is referred to as grading. A well-graded sediment is a (geologically) poorly sorted one, and a poorly graded sediment is a well-sorted one.

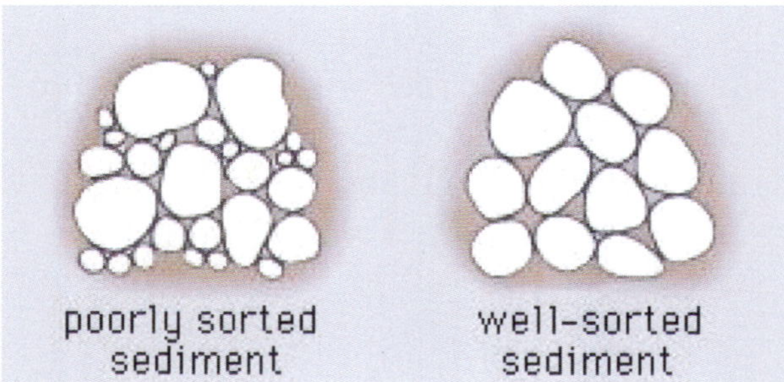

Total porosity encompasses all the void space, including those pores that are interconnected to the surface of the sample as well as those that are sealed off by natural cement or other obstructions. Thus the total porosity (ϕ_T) is

$$\varphi_T = 100\left(1 - \frac{Vol_G}{Vol_B}\right)\%,$$

where Vol_G is the volume of grains (and cement, if any) and Vol_B is the total bulk volume. Alternatively, one can calculate φ_T from the measured densities of the bulk rock and of the (mono) mineralic constituent. Thus,

$$\varphi_T = 100\left(1 - \frac{\rho_B}{\rho_G}\right)\%,$$

where ρ_B is the density of the bulk rock and ρ_G is the density of the grains (*i.e.*, the mineral, if the composition is monomineralogic and homogeneous). For example, if a sandstone has a ρ_B of 2.38 grams per cubic centimetre (g/cm^3) and is composed of quartz (SiO_2) grains having ρ_G of 2.65 g/cm^3, the total porosity is

$$\varphi_T = 100\left(1 - \frac{2.38}{2.65}\right) = 10.2\%.$$

Apparent (effective, or net) porosity is the proportion of void space that excludes the sealed-off pores. It thus measures the pore volume that is effectively interconnected and accessible to the surface of the sample, which is important when considering the storage and movement of subsurface fluids such as petroleum, groundwater, or contaminated fluids.

Physical Properties

Physical properties of rocks are of interest and utility in many fields of work, including geology, petrophysics, geophysics, materials science, geochemistry, and geotechnical engineering. The scale of investigation ranges from the molecular and crystalline up to terrestrial studies of the Earth and other planetary bodies. Geologists are interested in the radioactive age dating of rocks to reconstruct the origin of mineral deposits; seismologists formulate prospective earthquake predictions using premonitory physical or chemical changes; crystallographers study the synthesis of minerals with special optical or physical properties; exploration geophysicists investigate the variation of physical properties of subsurface rocks to make possible detection of natural resources such as oil and gas, geothermal energy, and ores of metals; geotechnical engineers examine the nature and behavior of the materials on, in, or of which such structures as buildings, dams, tunnels, bridges, and underground storage vaults are to be constructed; solid-state physicists study the magnetic, electrical, and mechanical properties of materials for electronic devices, computer components, or high-performance ceramics; and petroleum reservoir engineers analyze the response measured on well logs or in the processes of deep drilling at elevated temperature and pressure.

Since rocks are aggregates of mineral grains or crystals, their properties are determined in large part by the properties of their various constituent minerals. In a rock these general properties are determined by averaging the relative properties and sometimes orientations of the various grains or crystals. As a result, some properties that are anisotropic (*i.e.*, differ with direction) on a submicroscopic or crystalline scale are fairly isotropic for a large bulk volume of the rock. Many properties are also dependent on grain or crystal size, shape, and packing arrangement, the amount and distribution of void space, the presence of natural cements in sedimentary rocks, the temperature and pressure, and the type and amount of contained fluids (*e.g.*, water, petroleum, gases). Because many rocks exhibit a considerable range in these factors, the assignment of representative values for a particular property is often done using a statistical variation.

Some properties can vary considerably, depending on whether measured in situ (in place in the subsurface) or in the laboratory under simulated conditions. Electrical resistivity, for example, is highly dependent on the fluid content of the rock in situ and the temperature condition at the particular depth.

Density

Density varies significantly among different rock types because of differences in mineralogy and porosity. Knowledge of the distribution of underground rock densities can assist in interpreting subsurface geologic structure and rock type.

In strict usage, density is defined as the mass of a substance per unit volume; however, in common usage, it is taken to be the weight in air of a unit volume of a sample at a specific temperature. Weight is the force that gravitation exerts on a body (and thus varies with location), whereas mass (a measure of the matter in a body) is a fundamental property and is constant regardless of location. In routine density measurements of rocks, the sample weights are considered to be equivalent to their masses, because the discrepancy between weight and mass would result in less error on the computed density than the experimental errors introduced in the measurement of volume. Thus, density is often determined using weight rather than mass. Density should properly be reported in kilograms per cubic metre (kg/m^3), but is still often given in grams per cubic centimetre (g/cm^3).

Another property closely related to density is specific gravity. It is defined, as the ratio of the weight or mass in air of a unit volume of material at a stated temperature to the weight or mass in air of a unit volume of distilled water at the same temperature. Specific gravity is dimensionless (*i.e.*, has no units).

The bulk density of a rock is $\rho_B = W_G / V_B$, where W_G is the weight of grains (sedimentary rocks) or crystals (igneous and metamorphic rocks) and natural cements, if any, and V_B is the total volume of the grains or crystals plus the void (pore) space. The density can be dry if the pore space is empty, or it can be saturated if the pores are filled with fluid (*e.g.*, water), which is more typical of the subsurface (in situ) situation. If there is pore fluid present,

$$\text{(saturated)}\ \rho_B = \frac{W_G + W_{fl}}{V_B},$$

where W_{fl} is the weight of pore fluid. In terms of total porosity, saturated density is,

$$\text{(saturated)}\ \rho_B - \rho_G (1 - \varphi_T) + \rho_{fl} \varphi_T,$$

and thus $\varphi_T = \dfrac{\rho_G - \rho_B}{\rho_G - \rho_{fl}}$,

where ρ_{fl} is the density of the pore fluid. Density measurements for a given specimen involve the determination of any two of the following quantities: pore volume, bulk volume, or grain volume, along with the weight.

A useful way to assess the density of rocks is to make a histogram plot of the statistical range of a set of data. The representative value and its variation can be expressed as follows: (1) mean, the average value, (2) mode, the most common value (*i.e.*, the peak of the distribution curve), (3) median, the value of the middle sample of the data set (*i.e.*, the value at which half of the samples are below and half are above), and (4) standard deviation, a statistical measure of the spread of the data (plus and minus one standard deviation from the mean value includes about two-thirds of the data).

A compilation of dry bulk densities for various rock types found in the upper crust of the Earth is listed in the Table. A histogram plot of these data, giving the percent of the samples as a function of density is shown in figure below. The parameters given include (1) sample division, the range of density in one data column—*e.g.*, 0.036 g/cm³ for figure below, (2) number of samples, and (3) standard deviation. The small inset plot is the percentage of samples (on the vertical axis) that lie within the interval of the "mode - *x*" to the "mode + *x*," where *x* is the horizontal axis.

Dry Bulk Densities for Various Rock Types

rock type	number of samples	mean (grams per cubic cm)	standard deviation	mode (grams per cubic cm)	median (grams per cubic cm)
all rocks	1,647	2.73	0.26	2.65	2.86
andesite	197	2.65	0.13	2.58	2.66
basalt	323	2.74	0.47	2.88	2.87
diorite	68	2.86	0.12	2.89	2.87
dolerite (diabase)	224	2.89	0.13	2.96	2.90
gabbro	98	2.95	0.14	2.99	2.97
granite	334	2.66	0.06	2.66	2.66
quartz porphyry	76	2.62	0.06	2.60	2.62
rhyolite	94	2.51	0.13	2.60	2.49
syenite	93	2.70	0.10	2.67	2.68
trachyte	71	2.57	0.10	2.62	2.57
sandstone	107	2.22	0.23	2.22	2.22

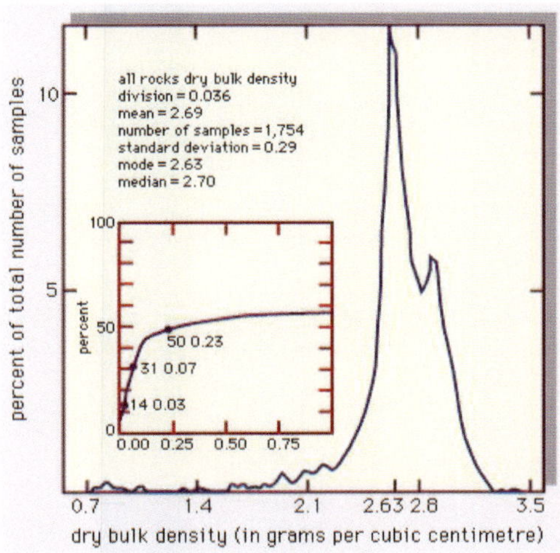

Figure: Dry bulk densities (distribution with density) for all rocks.

In Figure, the most common (modal) value of the distribution falls at 2.63 g/cm³, roughly the density of quartz, an abundant rock-forming mineral. Few density values for these upper crustal rocks lie above 3.3 g/cm³. A few fall well below the mode, even occasionally under 1 g/cm³. The reason for this is shown in Figure, which illustrates the density distributions for granite, basalt, and sandstone. Granite is an intrusive igneous rock with low porosity and a well-defined chemical (mineral) composition; its range of densities is narrow. Basalt is, in most cases, an extrusive igneous rock that can exhibit a large variation in porosity (because entrained gases leave voids called vesicles), and thus some highly porous samples can have low densities. Sandstone is a clastic sedimentary rock that can have a wide range of porosities depending on the degree of sorting, compaction, packing arrangement of grains, and cementation. The bulk density varies accordingly.

Other distribution plots of dry bulk densities are given in Figures, with a sample division of 0.036 g/cm^3 for Figures and of 0.828 percent for Figure B. The Table lists typical ranges of dry bulk densities for a variety of other rock types as prepared by the American geologists Gordon R. Johnson and Gary R. Olhoeft.

Typical Density Ranges for Some other Rock types

rock type	density (grams per cubic cm)
amphibolite	2.79–3.14
andesite glass	2.40–2.57
anhydrite	2.82–2.93
anorthosite	2.64–2.92
basalt glass	2.70–2.85
chalk	2.23
dolomite	2.72–2.84
dunite	2.98–3.76
eclogite	3.32–3.45
gneiss	2.59–2.84
granodiorite	2.67–2.78
limestone	1.55–2.75
marble	2.67–2.75
norite	2.72–3.02
peridotite	3.15–3.28
quartzite	2.65
rock salt	2.10–2.20
schist	2.73–3.19
shale	2.06–2.67
slate	2.72–2.84

The density of clastic sedimentary rocks increases as the rocks are progressively buried. This is because of the increase of overburden pressure, which causes compaction, and the progressive cementation with age. Both compaction and cementation decrease the porosity.

Representative densities for common rock-forming minerals (*i.e.*, ρG) and rocks (*i.e.*, ρ_B) are listed in the Table. The bulk densities for sedimentary rocks, which typically have variable porosity, are given as ranges of both dry ρ_B and (water-) saturated ρB.

Table 35: Typical Densities

material		ρ (g/cm³)	
minerals (approximately pure, no porosity)		ρ_G	
Halite (NaCl)		2.16	
Gypsum ($CaSO_4 \cdot 2H_2O$)		2.32	
Orthoclase feldspar ($KAlSi_3O_8$)		2.56	
Quartz (SiO_2)		2.65	
Calcite ($CaCO_3$)		2.71	
Anhydrite ($CaSO_4$)		2.9	
Olivine $(Mg,Fe)_2SiO_4$		3.3–3.6	
Hematite (Fe_2O_3)		5.26	
rocks (surface T and P)	typical porosity	dry ρ_B	saturated ρ_B
Loess		1.1	1.4–1.6
Silt		1.36	1.4–1.8
Rock salt		2.2	—
Shale	≳8%	1.9–2.4	2.1–2.5
Sandstone	~15%, variable	2.0–2.6	2.2–2.6
Limestone	~5%	1.9–2.6	2.1–2.7
Dolomite		2.5–2.8	2.6–2.8
Granite-diorite		2.6–2.9	
Gabbro-diabase		2.8–3.1	
Dunite-peridotite		3.1–3.4	
Granite; sial crust		2.67	
Basalt; sima crust		2.87	
Top of mantle		3.32	
other		ρ	
Water (0 °C)		0.99987	
20 °C		0.99823	
100 °C		0.95841	
Seawater (0 °C)		1.025	
Ice		0.92	
Oil		~0.8–0.9	
Methane gas (at 200 bars, $T \sim 80°$ C)		~0.12	

The pore-filling fluid is usually briny water, often indicative of the presence of seawater when the rock was being deposited or lithified. It should be noted that the bulk density is less than the grain density of the constituent mineral (or mineral assemblage), depending on the porosity. For example, sandstone (characteristically quartzose) has a typical dry bulk density of 2.0–2.6 g/cm³, with a porosity that can vary from low to more than 30 percent. The density of quartz itself is 2.65 g/cm³. If porosity were zero, the bulk density would equal the grain density.

Saturated bulk density is higher than dry bulk density, owing to the added presence of pore-filling fluid. The table also lists representative values for density of seawater, oil, and methane gas at a subsurface condition—pressure of 200 bars (one bar = 0.987 atmosphere, or 29.53 inches of mercury) and a temperature of about 80° C (176° F).

Mechanical Properties

Stress and Strain

When a stress σ (force per unit area) is applied to a material such as rock, the material experiences a change in dimension, volume, or shape. This change, or deformation, is called strain (ε). Stresses can be axial—*e.g.*,directional tension or simple compression—or shear (tangential), or all-sided (*e.g.*, hydrostatic compression). The terms stress and pressure are sometimes used interchangeably, but often stress refers to directional stress or shear stress and pressure (P) refers to hydrostatic compression. For small stresses, the strain is elastic (recoverable when the stress is removed and linearly proportional to the applied stress). For larger stresses and other conditions, the strain can be inelastic, or permanent.

Elastic Constants

In elastic deformation, there are various constants that relate the magnitude of the strain response to the applied stress. These elastic constants include the following:

1. Young's modulus (E) is the ratio of the applied stress to the fractional extension (or shortening) of the sample length parallel to the tension (or compression). The strain is the linear change in dimension divided by the original length.
2. Shear modulus (μ) is the ratio of the applied stress to the distortion (rotation) of a plane originally perpendicular to the applied shear stress; it is also termed the modulus of rigidity.
3. Bulk modulus (k) is the ratio of the confining pressure to the fractional reduction of volume in response to the applied hydrostatic pressure. The volume strain is the change in volume of the sample divided by the original volume. Bulk modulus is also termed the modulus of incompressibility.
4. Poisson's ratio (σp) is the ratio of lateral strain (perpendicular to an applied stress) to the longitudinal strain (parallel to applied stress).

For elastic and isotropic materials, the elastic constants are interrelated. For example,

$$\sigma_p = \frac{E}{2\mu} - 1;$$

$$E = 3k\left(1 - 2\sigma_p\right);$$

and,

$$\mu = \frac{3}{2}k\left(\frac{1 - 2\sigma_p}{1 + \sigma_p}\right).$$

The following are the common units of stress:

$$\begin{aligned} 1 \text{ bar} &= 10^6 \text{ dynes per square centimetre} \\ &= 10^5 \text{ newtons per square metr, or pascals (Pa)} \\ &= 0.1 \text{ megapascal } \left(\text{i.e., } 0.1 \times 10^6 \text{ Pa}\right). \end{aligned}$$

Thus $10 \text{ kilobars} = 1 \text{ gigapascal (i.e., } 10^9 \text{ Pa)}$.

Rock Mechanics

The study of deformation resulting from the strain of rocks in response to stresses is called rock mechanics. When the scale of the deformation is extended to large geologic structures in the crust of the Earth, the field of study is known as geotectonics.

The mechanisms and character of the deformation of rocks and Earth materials can be investigated through laboratory experiments, development of theoretical models based on the properties of materials, and study of deformed rocks and structures in the field. In the laboratory, one can simulate—either directly or by appropriate scaling of experimental parameters—several conditions. Two types of pressure may be simulated: confining (hydrostatic), due to burial under rock overburden, and internal (pore), due to pressure exerted by pore fluids contained in void space in the rock. Directed applied stress, such as compression, tension, and shear, is studied, as are the effects of increased temperature introduced with depth in the Earth's crust. The effects of the duration of time and the rate of applying stress (*i.e.,*loading) as a function of time are examined. Also, the role of fluids, particularly if they are chemically active, is investigated.

Some simple apparatuses for deforming rocks are designed for biaxial stress application: a directed (uniaxial) compression is applied while a confining pressure is exerted (by pressurized fluid) around the cylindrical specimen. This simulates deformation at depth within the Earth. An independent internal pore-fluid pressure also can be exerted. The rock specimen can be jacketed with a thin, impermeable sleeve (*e.g.,* rubber or copper) to separate the external pressure medium from the internal pore fluids (if any). The specimen is typically a few centimetres in dimension.

Another apparatus for exerting high pressure on a sample was designed in 1968 by Akira Sawaoka, Naoto Kawai, and Robert Carmichael to give hydrostatic confining pressures up to 12 kilobars (1.2 gigapascal), additional directed stress, and temperatures up to a few hundred degrees Celsius. The specimen is positioned on the baseplate; the pressure is applied by driving in pistons with a hydraulic press. The end caps can be locked down to hold the pressure for time experiments and to make the device portable.

Apparatuses have been developed, typically using multianvil designs, which extend the range of static experimental conditions—at least for small specimens and limited times—to pressures as high as about 1,700 kilobars and temperatures of about 2,000° C. Such work has been pioneered by researchers such as Peter M. Bell and Ho-Kwang Mao, who conducted studies at the Geophysical Laboratory of the Carnegie Institution in Washington, D.C. Using dynamic techniques (*i.e.,* shock from explosive impact generated by gun-type designs), even higher pressures up to 7,000 kilobars (700 gigapascal)—which is nearly twice the pressure at the centre of the Earth and seven million times greater than the atmospheric pressure at the Earth's surface—can be produced for very short times. A leading figure in such ultrapressure work is A. Sawaoka at the Tokyo Institute of Technology.

In the upper crust of the Earth, hydrostatic pressure increases at the rate of about 320 bars per kilometre, and temperature increases at a typical rate of 20°–40°C per kilometre, depending on recent crustal geologic history. Additional directed stress, as can be generated by large-scale crustal deformation (tectonism), can range up to 1 to 2 kilobars. This is approximately equal to the ultimate strength (before fracture) of solid crystalline rock at surface temperature and pressure. The stress released in a single major earthquake—a shift on a fault plane—is about 50–150 bars.

In studying the deformation of rocks one can start with the assumption of ideal behaviour: elastic strain and homogeneous and isotropic stress and strain. In reality, on a microscopic scale there are grains and pores in sediments and a fabric of crystals in igneous and metamorphic rocks. On a large scale, rock bodies exhibit physical and chemical variations and structural features. Furthermore, conditions such as extended length of time, confining pressure, and subsurface fluids affect the rates of change of deformation. Figure below shows the generalized transition from brittle fracture through faulting to plastic-flow deformation in response to applied compressional stress and the progressive increase of confining pressure.

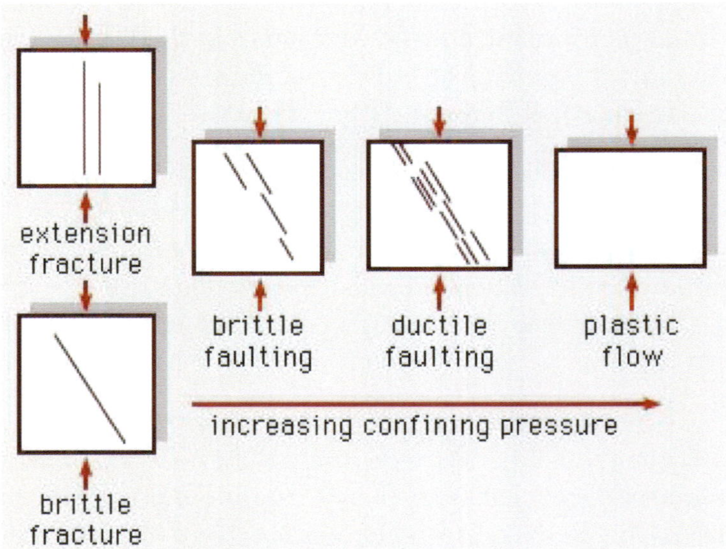

Figure: Deformation as affected by increased confining pressure.

Stress-Strain Relationships

The deformation of materials is characterized by stress-strain relations. For elastic-behaviour materials, the strain is proportional to the load (*i.e.*,the applied stress). The strain is immediate with stress and is reversible (recoverable) up to the yield point stress, beyond which permanent strain results. For viscous material, there is laminar (slow, smooth, parallel) flow; one must exert a force to maintain motion because of internal frictional resistance to flow, called the viscosity. Viscosity varies with the applied stress, strain rate, and temperature. In plastic behavior, the material strains continuously (but still has strength) after the yield point stress is reached; however, beyond this point there is some permanent deformation. In elasticoviscous deformation, there is combined elastic and viscous behavior. The material yields continuously (viscously) for a constant applied load. An example of such behavior is creep, a slow, permanent, and continuous deformation occurring under constant load over a long time in such materials as crystals, ice, soil and sediment, and rocks at depth. In firm viscous behavior, the material is essentially solid but the strain is not immediate with application of stress; rather, it is taken up and released exponentially. A plasticoviscous material exhibits elastic behavior for initial stress (as in plastic behavior), but after the yield point stress is reached, it flows like a viscous fluid.

Some representative values of elastic constants and properties are listed in Table . The coefficient of viscosity (η) is the ratio of applied stress to the rate of straining (change of strain with time). It is measured in units of poise; one poise equals one dyne-second per square centimetre.

Some Typical Values of Elastic Constants and Properties

	elastic constants (at room temperature and pressure)	
material	Young's modulus (in 10^6bars)	shear modulus (in 10^6bars)
ice	0.1	0.03
shale	0.2–0.3	0.15
limestone	0.4–0.7	0.22–0.26
granite	0.3–0.6	0.2
basalt	0.7–0.9	0.3
steel	2.1	0.83
material	temperature (degrees Celsius)	coefficient of viscosity (poises)
lava (Mount Vesuvius)	1,100 1,400	28,300 250
lava (Oshima, Japan)	1,038 1,125	230,000 5,600
andesite lava	1,400	150–1,500
material	compressive strength (at room temperature and pressure, in kilobars)	
shale	0.8–1.8	
sandstone	0.5–2	
limestone	1–2	
granite	1.7–2.5	
basalt	1–3.4	

Rheology is the study of the flow deformation of materials. The concept of rheidity refers to the capacity of a material to flow, arbitrarily defined as the time required with a shear stress applied for the viscous strain to be 1,000 times greater than the elastic strain. It is thus a measure of the threshold of fluidlike behaviour. Although such behaviour depends on temperature, relative comparisons can be made. Some representative values of rheidity times are given in the Table.

Rheidity Threshold of Fluidlike Deformation

material	approximate time
ice (e.g., glacier)	2 weeks
gypsum	1 year
rock salt (e.g., saltdome)	10–20 years
serpentine (a mafic silicate mineral)	10,000 years

Typical stress-strain (deformation) curves for rock materials are shown in Figure. The stress σ, compression in the figure, is force per unit area. The strain ε is fractional shortening of the specimen parallel to the applied compression; it is given here in percent. The brittle material behaves elastically nearly until the point of fracture (denoted *X*), whereas the ductile (plastically deformable) material is elastic up to the yield point but then has a range of plastic deformation before fracturing. The ability to undergo large permanent deformation before fracture is called ductility. For plastic deformation, the flow mechanisms are intracrystalline (slip and twinning within crystal grains), intercrystalline motion by crushing and fracture (cataclasis), and recrystallization by solutioning or solid diffusion.

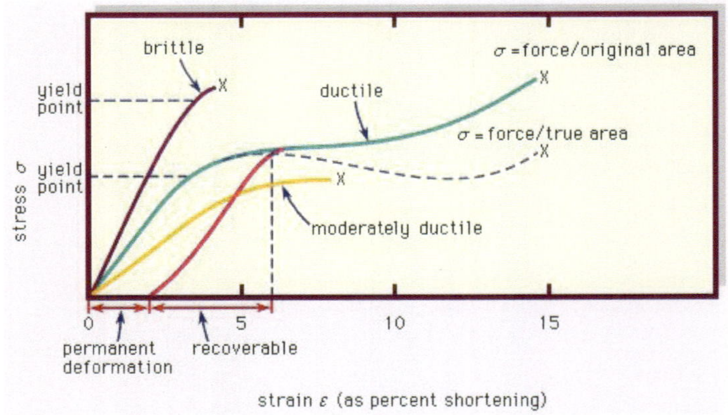

Figure: Typical stress-strain curves for rock materials. Each *X* represents the point of fracture for the corresponding material.

If the applied stress is removed while a ductile material is in the plastic range, part of the strain is recoverable (elastically), but there is permanent deformation. The ultimate strength is the highest point (stress) on a stress-strain curve, often occurring at fracture (which is the complete loss of cohesion). The strength of a material is its resistance to failure (destruction of structure) by flow or fracture; it is a measure of the stress required to deform a body. Typical compressive strengths (the stress required to cause failure under compression) are given in the table.

Effect of Environmental Conditions

The behaviour and mechanical properties of rocks depend on a number of environmental conditions. (1) Confining pressure increases the elasticity, strength (*e.g.,* yield point and ultimate fracture stress), and ductility. (2) Internal pore-fluid pressure reduces the effective stress acting on the sample, thus reducing the strength and ductility. The effective, or net, confining pressure is the external hydrostatic pressure minus the internal pore-fluid pressure. (3) Temperature lowers the strength, enhances ductility, and may enhance recrystallization. (4) Fluid solutions can enhance deformation, creep, and recrystallization. (5) Time is an influential factor as well. (6) The rate of loading (*i.e.,* the rate at which stress is applied) influences mechanical properties. (7) Compaction, as would occur with burial to depth, reduces the volume of pore space for sedimentary rocks and the crack porosity for crystalline rocks.

Rocks, which are typically brittle at the Earth's surface, can undergo ductile deformation when buried and subjected to increased confining pressure and temperature for long periods of time. If stress exceeds their strength or if they are not sufficiently ductile, they will fail by fracture—as a crystal, within a bed or rock, on an earthquake fault zone, and so on—whereas with ductility they can flow and fold.

Some strengths for various rock types under different temperatures and confining pressures are listed in the table. The plastic yield strength here is the stress at a 2 percent strain; the ultimate strength, as stated above, is the highest point (stress) on the stress-strain curve.

Rock strengths, with varying temperature and pressure				
rock type	temperature (°C)	confining pressure (kilobars)	plastic yield strength (kilobars)	ultimate strength (kilobars)
granite	500	5	10	11.5
	800	5	5	6
gabbro	500	5	4	8
peridotite	500	5	8	9
	800	5	5.5	8
basalt	500	5	8	10
	800	5	2	2.5
marble	24	2	2.5	5.5
	500	3	1	2
limestone	24	2	4.5	5.5
	500	3	2.5	3
dolomite	24	2	6	7
	500	5	4	6.5
shale	24	2	1.5	2.5
rock salt	24	1	0.5	1

An increase in confining pressure causes brittle fracture to become shear slippage and eventually causes flow (ductile) behaviour. This transition is also aided by higher temperature, decreased internal pore-fluid pressure, and slower strain rate.

The Table gives the values of some elastic constants—bulk modulus (k), Young's modulus (E), shear modulus (μ), and Poisson's ratio (σp)—at room pressure (1 bar) and high confining pressure (3,000 bars). The values for clastic sedimentary rocks would be particularly variable.

Variation of Some Elastic Constants (in 10^6 Bars) with Rock Type and Confining Pressure

at pressure = 1 bar				
rock type	bulk modulus	Young's modulus	shear modulus	Poisson's ratio
granite	0.1	0.3	0.2	0.05
gabbro	0.3	0.9	0.6	0.1
dunite	1.1	1.5	0.5	0.3
obsidian	0.4	0.7	0.3	0.08
basalt	0.5	0.8	0.3	0.23
gneiss	0.1	0.2	0.1	0.05
marble	0.1	0.4	0.2	0.1
quartzite				
sandstone	0.07	0.2	0.08	0.1
shale	0.04	0.1	0.05	0.04
limestone	0.8	0.6	0.2	0.3

at pressure = 1 bar				
rock type	bulk modulus	Young's modulus	shear modulus	Poisson's ratio
granite	0.1	0.3	0.2	0.05
gabbro	0.3	0.9	0.6	0.1
at pressure = 3,000 bars				
rock type	bulk modulus	Young's modulus	shear modulus	Poisson's ratio
granite	0.5	0.6	0.4	0.25
gabbro	0.9	0.8	0.5	0.2
dunite	1.2	1.7	0.7	0.27
obsidian				
basalt	0.8	1.2	0.4	0.25
gneiss	0.5	0.7	0.3	
marble	0.8	0.7	0.3	0.3
quartzite	0.5	1.0	0.4	0.07
sandstone				
shale				
limestone				

Thermal Properties

Heat flow (or flux), q, in the Earth's crust or in rock as a building material, is the product of the temperature gradient (change in temperature per unit distance) and the material's thermal conductivity (k, the heat flowacross a surface per unit area per unit time when a temperature difference exists in unit length perpendicular to the surface). Thus,

$$q = \frac{dT}{dz} \times k.$$

The units of the terms in this equation are given below, expressed first in the centimetre-gram-second (cgs) system and then in the International System of Units (SI) system, with the conversion factor from the first to the second given between them.

q, *heat flow*:

calories per square centime per second $(cal / cm^2 \cdot sec)$ $\times \frac{1}{23.9} \times 10^{-6}$ = *watta per square metre* (W / m^2)

$\frac{dT}{dz}$, *temperature gradient*:

degrees Celsius kilometre $(^\circ C / km$, *practical unit for Earth*$)$ $\times 10^{-3}$ = *degrees Celsius per meter* $(^\circ C / m)$

k, *thermal conductivity*:

calories per square centimetre per second per degree Celsius $(cal / cm^2 \cdot sec \cdot {}^\circ C)$ $\times \frac{1}{23.9} \times 10^{-3}$ *watts per metre per degree Celsius* $(W / m \cdot {}^\circ C)$

Thermal Conductivity

Thermal conductivity can be determined in the laboratory or in situ, as in a borehole or deep well, by turning on a heating element and measuring the rise in temperature with time. It depends on several factors: (1) chemical composition of the rock (*i.e.*, mineral content), (2) fluid content (type and degree of saturation of the pore space); the presence of water increases the thermal conductivity (*i.e.*, enhances the flow of heat), (3) pressure (a high pressure increases the thermal conductivity by closing cracks which inhibit heat flow), (4) temperature, and (5) isotropy and homogeneity of the rock.

Typical values of thermal conductivities of rock materials are given in the Table. For crystalline silicate rocks—the dominant rocks of the "basement" crustal rocks—the lower values are typical of ones rich in magnesium and iron (*e.g.*, basalt and gabbro) and the higher values are typical of those rich in silica (quartz) and alumina (*e.g.*, granite). These values result because the thermal conductivity of quartz is relatively high, while that for feldspars is low.

Typical values of thermal conductivity (in 0.001 calories per centimetre per second per degree Celsius)		
material	at 20° C	at 200° C
typical rocks	4–10	
granite	7.8	6.6
gneiss		
(perpendicular to banding)	5.9	5.5 (100° C)
(parallel to banding)	8.2	7.4 (100° C)
gabbro	5.1	5.0
basalt	4.0	4.0
dunite	12.0	8.1
marble	7.3	5.2
quartzite	15.0	9.0
limestone	6.0	
one sandstone		
(dry)	4.4	
(saturated)	5.4	
shale	3–4	
rock salt	12.8	
sand		
(dry)	0.65	
(30% water)	3.94	
water	1.34 (0° C)	1.6 (80° C)
ice	5.3 (0° C)	9.6 (−130° C)
magnetite	12.6	
quartz	20.0	
feldspars	5.0	

Thermal Expansion

The change in dimension—linear or volumetric—of a rock specimen with temperature is expressed in terms of a coefficient of thermal expansion. This is given as the ratio of dimension change (*e.g.*, change in volume) to the original dimension (volume, V) per unit of temperature (T) change:

$$\frac{1}{V}\frac{\Delta V}{\Delta T}.$$

Most rocks have a volume-expansion coefficient in the range of 15–33 × 10^{-6} per degree Celsius under ordinary conditions. Quartz-rich rocks have relatively high values because of the higher volume expansion coefficient of quartz. Thermal-expansion coefficients increase with temperature. Table lists some linear-expansion coefficients,

Thermal expansion of rocks	
rock type	linear-expansion coefficient (in 10^{-6} per degree Celsius)
granite and rhyolite	8 ± 3
andesite and diorite	7 ± 2
basalt, gabbro, and diabase	5.4 ± 1
sandstone	10 ± 2
limestone	8 ± 4
marble	7 ± 2
slate	9 ± 1

$$\frac{1}{L}\frac{\Delta L}{\Delta T},$$

where L represents length. All data are based on at least three samples.

Radioactive Heat Generation

The spontaneous decay (partial disintegration) of the nuclei of radioactive elements provides decay particles and energy. The energy, composed of emission kinetic energy and radiation, is converted to heat; it has been an important factor in affecting the temperature gradient and thermal evolution of the Earth. Deep-seated elevated temperatures provide the heat that causes rock to deform plastically and to move, thus generating to a large extent the processes of plate tectonics—plate motions, seafloor spreading, continental drift, and subduction—and most earthquakes and volcanism.

Some elements, or their isotopes (nuclear species with the same atomic number but different mass numbers), decay with time. These include elements with an atomic number greater than 83—of which the most important are uranium-235, uranium-238, and thorium-232—and a few with a lower atomic number, such as potassium-40.

The heat generated within rocks depends on the types and abundances of the radioactive elements and their host minerals. Such heat production, A, is given in calories per cubic centimetre per second, or 1 calorie per gram per year = 4.186 × 10^7 ergs per gram per year = 1.327 ergs per gram per second. The rate of radioactive decay, statistically an exponential process, is given by the half-life, $t1/2$. The half-life is the time required for half the original radioactive atoms to decay for a particular isotope.

Some radioactive decay series are listed in the Table. The isotopic abundance is the percent of the natural element that exists as that particular radioactive isotope; for example, 99.28 percent of natural uranium is U-238, and 100 percent of thorium is the radioactive Th-232. The final product is the end result of the process (usually multistage) of disintegration. The Table gives the heat productivities of radioactive elements and rock types as reported by George D. Garland. For the rocks, the typical content is given for uranium and thorium (in parts per million [ppm] of weight) and for potassium (in weight percent). The heat production of natural uranium is close to that for the isotope U-238, since almost all natural uranium is of that isotopic species.

Heat Productivities

isotope	heat productivity, A (calories per gram per year)			
U-235	4.29			
U-238	0.71			
natural uranium	0.73			
Th-232	0.20			
K-40	0.22			
natural potassium	$27(10^{-6})$			
Rb-87	$130(10^{-6})$			
natural rubidium	$36(10^{-6})$			
major rock province	concentration			heat productivity, A
	U (ppm)	Th (ppm)	K (%)	($\times 10^{-13}$ calories per cubic centimetre per second)
oceanic crust	0.42	1.68	0.69	0.71
continental shield crust (old)	1.00	4.00	1.63	1.67
continental upper crust (young)	1.32	5.28	2.15	2.20
Some radioactive decay series				
*half-life for K-40 as a whole is $1.25(10^9)$ years.				
element	radioactive isotope	final product	isotopic abundance (%)	half-life (in 10^9years)
uranium	U-235	Pb-207	0.72	0.7
	U-238	Pb-206	99.28	4.5
thorium	Th-232	Pb-208	100.0	14.0
potassium	K-40	(89%) Ca-40	0.01	1.4*
		(11%) Argon-40		11.9*
rubidium	Rb-87	Sr-87	27.8	48.8

*half-life for K-40 as a whole is 1.25(109) years.

The radioactive elements are more concentrated in the continental upper-crust rocks that are rich in quartz (*i.e.*, felsic, or less mafic). This results because these rocks are differentiated by partial melting of the upper-mantle and oceanic-crust rock. The radioactive elements tend to be preferentially driven off from these rocks for geochemical reasons. A compilation of heat productivities of various rock types is given in the table.

Heat productivities of various rocks				
rock type	abundances			
	U ppm	Th ppm	Rb ppm	K %
granite	3.4	50	220	4.45
andesite	1.9	6.4	67	2.35
oceanic basalt	0.5	0.9	9	0.43
peridotite	0.005	0.01	0.063	0.001
average upper-continental crust	2.5	10.5	110	2.7
average continental crust	1.0	2.5	50	1.25
rock type	heat production			total A (in 10^{-6} calories per gram per year)
	from U	from Th	from K	
granite	2.52	9.95	1.16	13.63
andesite	1.41	1.27	0.61	3.29
oceanic basalt	0.37	0.18	0.11	0.66
peridotite	0.0037	0.002	0.0003	0.006
average upper-continental crust	1.85	2.09	0.7	4.64
average continental crust	0.74	0.5	0.33	1.56

Electrical Properties

The electrical nature of a material is characterized by its conductivity (or, inversely, its resistivity) and its dielectric constant, and coefficients that indicate the rates of change of these with temperature, frequency at which measurement is made, and so on. For rocks with a range of chemical composition as well as variable physical properties of porosity and fluid content, the values of electrical properties can vary widely.

Resistance (R) is defined as being one ohm when a potential difference (voltage; *V*) across a specimen of one volt magnitude produces a current (i) of one ampere; that is, $V = Ri$. The electrical resistivity (ρ) is an intrinsicproperty of the material. In other words, it is inherent and not dependent on sample size or current path. It is related to resistance by $R = \rho L / A$ where L is the length of specimen, A is the cross-sectional area of specimen, and units of ρ are ohm-centimetre; 1 ohm-centimetre equals 0.01 ohm-metre. The conductivity (σ) is equal to $1/\rho$ ohm $^{-1}$ centimetre^{-1}(or termed mhos/cm). In SI units, it is given in mhos/metre, or siemens/metre.

Some representative values of electrical resistivity for rocks and other materials are listed in the Table. Materials that are generally considered as "good" conductors have a resistivity of $10^{-5} - 10$ ohm-centimetre ($10^{-7} - 10^{-1}$ ohm-metre) and a conductivity of $10 - 10^{7}$ mhos/metre. Those that are classified as intermediate conductors have a resistivity of 100–10^9 ohm-centimetre (1–10^7 ohm-metre) and a conductivity of $10^{-7} - 1$ mhos/metre. "Poor" conductors, also known as insulators, have a resistivity of $10^{10} - 10^{17}$ ohm-centimetre ($10^{8} - 10^{15}$ ohm-metre) and a conductivity of $10^{-15} - 10^{-8}$. Seawater is a much better conductor (*i.e.*, it has lower resistivity) than fresh water owing to its higher content of dissolved salts; dry rock is very resistive. In the subsurface, pores are typically filled to some degree by fluids. The resistivity of materials has a wide range—copper is, for example, different from quartz by 22 orders of magnitude.

Typical Resistivities

material		resistivity (ohm-centimetre)
seawater (18° C)		21
uncontaminated surface water		$2(10^4)$
distilled water		$0.2–1(10^6)$
water (4° C)		$9(10^6)$
ice		$3(10^8)$
rocks in situ		
sedimentary	clay, soft shale	$100–5(10^3)$
	hard shale	$7–50(10^3)$
	sand	$5–40(10^3)$
	sandstone	$(10^4)–(10^5)$
	glacial moraine	$1–500(10^3)$
	porous limestone	$1–30(10^4)$
	dense limestone	$>(10^6)$
	rock salt	$(10^8)–(10^9)$
igneous		$5(10^4)–(10^8)$
metamorphic		$5(10^4)–5(10^9)$
rocks in laboratory		
dry granite		10^{12}
minerals		
copper (18° C)		$1.7(10^{-6})$
graphite		$5–500(10^{-4})$
pyrrhotite		0.1–0.6
magnetite crystals		0.6–0.8
pyrite ore		$1–(10^5)$
magnetite ore		$(10^2)–5(10^5)$
chromite ore		$>10^6$
quartz (18° C)		$(10^{14})–(10^{16})$

For high-frequency alternating currents, the electrical response of a rock is governed in part by the dielectric constant, ε. This is the capacity of the rock to store electric charge; it is a measure of polarizability in an electric field. In cgs units, the dielectric constant is 1.0 in a vacuum. In SI units, it is given in farads per metre or in terms of the ratio of specific capacity of the material to specific capacity of vacuum (which is 8.85×10^{-12} farads per metre). The dielectric constant is a function of temperature, and of frequency, for those frequencies well above 100 hertz (cycles per second).

Electrical conduction occurs in rocks by (1) fluid conduction—*i.e.,*electrolytic conduction by ionic transfer in briny pore water—and (2) metallic and semiconductor (*e.g.,* some sulfide ores) electron conduction. If the rock has any porosity and contained fluid, the fluid typically dominates the conductivity response. The rock conductivity depends on the conductivity of the fluid (and its chemical composition), degree of fluid saturation, porosity and permeability, and temperature. If rocks lose water, as with compaction of clastic sedimentary rocks at depth, their resistivity typically increases.

Magnetic Properties

The magnetic properties of rocks arise from the magnetic properties of the constituent mineral grains and crystals. Typically, only a small fraction of the rock consists of magnetic minerals. It is this small portion of grains that determines the magnetic properties and magnetization of the rock as a whole, with two results: (1) the magnetic properties of a given rock may vary widely within a given rock body or structure, depending on chemical inhomogeneities, depositional or crystallization conditions, and what happens to the rock after formation; and (2) rocks that share the same lithology (type and name) need not necessarily share the same magnetic characteristics. Lithologic classifications are usually based on the abundance of dominant silicate minerals, but the magnetization is determined by the minor fraction of such magnetic mineral grains as iron oxides. The major rock-forming magnetic minerals are iron oxides and sulfides.

Although the magnetic properties of rocks sharing the same classification may vary from rock to rock, general magnetic properties do nonetheless usually depend on rock type and overall composition. The magnetic properties of a particular rock can be quite well understood provided one has specific information about the magnetic properties of crystalline materials and minerals, as well as about how those properties are affected by such factors as temperature, pressure, chemical composition, and the size of the grains. Understanding is further enhanced by information about how the properties of typical rocks are dependent on the geologic environment and how they vary with different conditions.

Applications of the Study of Rock Magnetization

An understanding of rock magnetization is important in at least three different areas: prospecting, geology, and materials science. In magnetic prospecting, one is interested in mapping the depth, size, type, and inferred composition of buried rocks. The prospecting, which may be done from ground surface, ship, or aircraft, provides an important first step in exploring buried geologic structures and may, for example, help identify favourable locations for oil, natural gas, and economic mineral deposits.

Rock magnetization has traditionally played an important role in geology. Paleomagnetic work seeks to determine the remanent magnetization and thereby ascertain the character of the Earth's field when certain rocks were formed. The results of such research have important ramifications in stratigraphic correlation, age dating, and reconstructing past movements of the Earth's crust. Indeed, magnetic surveys of the oceanic crust provided for the first time the quantitative evidence needed to cogently demonstrate that segments of the crust had undergone large-scale lateral displacements over geologic time, thereby corroboratingthe concepts of continental drift and seafloor spreading, both of which are fundamental to the theory of plate tectonics (see plate tectonics).

The understanding of magnetization is increasingly important in materials science as well. The design and manufacture of efficient memory cores, magnetic tapes, and permanent magnets increasingly rely on the ability to create materials having desired magnetic properties.

Basic Types of Magnetization

There are six basic types of magnetization: (1) diamagnetism, (2) paramagnetism, (3) ferromagnetism, (4) antiferromagnetism, (5) ferrimagnetism, and (6) superparamagnetism.

Diamagnetism arises from the orbiting electrons surrounding each atomic nucleus. When an external magnetic field is applied, the orbits are shifted in such a way that the atoms set up their own magnetic field in opposition to the applied field. In other words, the induced diamagnetic field opposes the external field. Diamagnetism is present in all materials, is weak, and exists only in the presence of an applied field. The propensity of a substance for being magnetized in an external field is called its susceptibility (k) and it is defined as J/H, where J is the magnetization (intensity) per unit volume and H is the strength of the applied field. Since the induced field always opposes the applied field, the sign of diamagnetic susceptibility is negative. The susceptibility of a diamagnetic substance is on the order of -10^{-6} electromagnetic units per cubic centimetre (emu/cm^3). It is sometimes denoted κ for susceptibility per unit mass of material.

Paramagnetism results from the electron spin of unpaired electrons. An electron has a magnetic dipole moment—which is to say that it behaves like a tiny bar magnet—and so when a group of electrons is placed in a magnetic field, the dipole moments tend to line up with the field. The effect augments the net magnetization in the direction of the applied field. Like diamagnetism, paramagnetism is weak and exists only in the presence of an applied field, but since the effect enhances the applied field, the sign of the paramagnetic susceptibility is always positive. The susceptibility of a paramagnetic substance is on the order of 10^{-4} to 10^{-6}emu/cm^3.

Ferromagnetism also exists because of the magnetic properties of the electron. Unlike paramagnetism, however, ferromagnetism can occur even if no external field is applied. The magnetic dipole moments of the atoms spontaneously line up with one another because it is energetically favourable for them to do so. A remanent magnetization can be retained. Complete alignment of the dipole moments would take place only at a temperature of absolute zero (0 kelvin, or -273.15°C). Above absolute zero, thermal motions begin to disorder the magnetic moments. At a temperature called the Curie temperature, which varies from material to material, the thermally induced disorder overcomes the alignment, and the ferromagnetic properties of the substance disappear. The susceptibility of ferromagnetic materials is large and positive. It is on the order of 10 to 10^4 emu/cm^3. Only a few materials—iron, cobalt, and nickel—are ferromagnetic in the strict sense of the word and have a strong residual magnetization. In general usage, particularly in engineering, the term ferromagnetic is frequently applied to any material that is appreciably magnetic.

Antiferromagnetism occurs when the dipole moments of the atoms in a material assume an antiparallel arrangement in the absence of an applied field. The result is that the sample has no net magnetization. The strength of the susceptibility is comparable to that of paramagnetic materials. Above a temperature called the Néel temperature, thermal motions destroy the antiparallel arrangement, and the material then becomes paramagnetic. Spin-canted (anti)ferromagnetism is a special condition which occurs when antiparallel magnetic moments are deflected from the antiferromagnetic plane, resulting in a weak net magnetism. Hematite $(\alpha - Fe_2O_3)$ is such a material.

Ferrimagnetism is an antiparallel alignment of atomic dipole moments which does yield an appreciable net magnetization resulting from unequal moments of the magnetic sublattices. Remanent magnetization is detectable. Above the Curie temperature the substance becomes paramagnetic. Magnetite (Fe_3O_4), which is the most magnetic common mineral, is a ferrimagnetic substance.

Superparamagnetism occurs in materials having grains so small (about 100 angstroms) that any cooperative alignment of dipole moments is overcome by thermal energy.

Types of Remanent Magnetization

Rocks and minerals may retain magnetization after the removal of an externally applied field, thereby becoming permanent weak magnets. This property is known as remanent magnetization and is manifested in different forms, depending on the magnetic properties of the rocks and minerals and their geologic origin and history. Delineated below are the kinds of remanent magnetization frequently observed.

CRM (chemical, or crystallization, remanent magnetization) can be induced after a crystal is formed and undergoes one of a number of physicochemical changes, such as oxidation or reduction, a phase change, dehydration, recrystallization, or precipitation of natural cements. The induction, which is particularly important in some (red) sediments and metamorphic rocks, typically takes place at constant temperature in the Earth's magnetic field.

DRM (depositional, or detrital, remanent magnetization) is formed in clastic sediments when fine particles are deposited on the floor of a body of water. Marine sediments, lake sediments, and some clays can acquire DRM. The Earth's magnetic field aligns the grains, yielding a preferred direction of magnetization.

IRM (isothermal remanent magnetization) results from the application of a magnetic field at a constant (isothermal) temperature, often room temperature.

NRM (natural remanent magnetization) is the magnetization detected in a geologic in situ condition. The NRM of a substance may, of course, be a combination of any of the other remanent magnetizations described here.

PRM (pressure remanent, or piezoremanent, magnetization) arises when a material undergoes mechanical deformation while in a magnetic field. The process of deformation may result from hydrostatic pressure, shock impact (as produced by a meteorite striking the Earth's surface), or directed tectonic stress. There are magnetization changes with stress in the elastic range, but the most pronounced effects occur with plastic deformation when the structure of the magnetic minerals is irreversibly changed.

TRM (thermoremanent magnetization) occurs when a substance is cooled, in the presence of a magnetic field, from above its Curie temperature to below that temperature. This form of magnetization is generally the most important, because it is stable and widespread, occurring in igneous and sedimentary rocks. TRM also can occur when dealing exclusively with temperatures below the Curie temperature. In PTRM (partial thermoremanent magnetization) a sample is cooled from a temperature below the Curie point to yet a lower temperature.

VRM (viscous remanent magnetization) results from thermal agitation. It is acquired slowly over time at low temperatures and in the Earth's magnetic field. The effect is weak and unstable but is present in most rocks.

Hysteresis and Magnetic Susceptibility

The concept of hysteresis is fundamental when describing and comparing the magnetic properties of rocks. Hysteresis is the variation of magnetization with applied field and illustrates the ability of a material to retain its magnetization, even after an applied field is removed. Figure below illustrates this phenomenon in the form of a plot of magnetization (J) versus applied field (H_{ex}). J_s is the saturation (or "spontaneous") magnetization when all the magnetic moments are aligned in their configuration of maximum order. It is temperature-dependent, reaching zero at the Curie temperature. $J_{r,sat}$ is the remanent magnetization that remains when a saturating (large) applied field is removed, and J_r is the residual magnetization left by some process apart from IRM saturation, as, for example, TRM. H_c is the coercive field (or force) that is required to reduce $J_{r,sat}$ to zero, and $H_{c,r}$ is the field required to reduce J_r to zero.

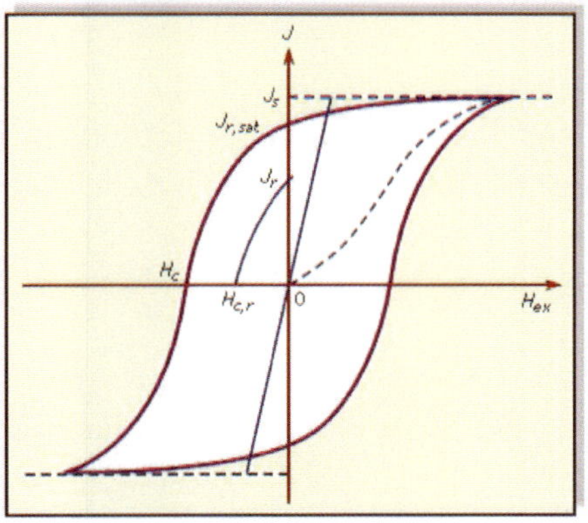

In figure, General magnetic hysteresis curve, showing magnetization (J) as a function of the external field (*Hex*). J_s is the saturation (or "spontaneous") magnetization; $J_{r,sat}$ is the remanent magnetization that remains after a saturating applied field is removed; J_r is the residual magnetization left by some magnetization process other than IRM saturation; H_c is the coercive field; and $H_{c,r}$ is the field necessary to reduce J_r to zero.

Magnetic susceptibility is a parameter of considerable diagnostic and interpretational use in the study of rocks. This is true whether an investigation is being conducted in the laboratory or magnetic fields over a terrain are being studied to deduce the structure and lithologic character of buried rock bodies. Susceptibility for a rock type can vary widely, depending on magnetic mineralogy, grain size and shape, and the relative magnitude of remanent magnetization present, in addition to the induced magnetization from the Earth's weak field. The latter is given as $J_{induced} = kH_{ex}$, where k is the (true) magnetic susceptibility and *Hex* is the external (*i.e.,* the Earth's) magnetic field. If there is an additional remanent magnetization with its ratio (Q_n) to induced magnetization being given by,

$$Q_n = \frac{J_{induced}}{kH_{ex}},$$

then the total magnetization is,

$$\begin{aligned} J &= J_{induced} + J_{remanent} \\ &= kH_{ex} + Q_n kH_{ex} \\ &= k(1+Q_n)H_{ex} \\ &= k_{app}H_{ex} \end{aligned}$$

where k_{app}, the "apparent" magnetic susceptibility, is $k(1+Q_n)$.

Magnetic Minerals and Magnetic Properties of Rocks

The major rock-forming magnetic minerals are the following iron oxides: the titanomagnetite series, $xFe_2TiO_4 \cdot (1 - x)Fe_3O_4$, where Fe_3O_4 is magnetite, the most magnetic mineral; the ilmeno-hematite series $yFeTiO_3 \cdot (1 - y)Fe_2O_3$, where $\alpha - Fe_2O_3$ (in its rhombohedral structure) is hematite; maghemite, $\gamma - Fe_2O_3$ (in which some iron atoms are missing in the hematite structure); and limonite (hydrous iron oxides). They also include sulfides—namely, the pyrrhotite series, $yFeS \cdot (1 - y)Fe_{1-x}S$.

The Table gives some typical values of the apparent susceptibility for various rock types, which usually include some remanent as well as induced magnetization. Values are higher for mafic igneous rocks, especially as the content of magnetite increases.

Approximate "Apparent Susceptibilities" for Rock Types

rock	apparent magnetic susceptibility (electromagnetic units per cubic centimetre)
Source: From T. Nagata (ed.), Rock Magnetism, Maruzen Co., Tokyo (1961).	
iron ores	over 0.1
basalt	10^{-2}
andesite	10^{-3}
dacite	10^{-4}
metamorphic rocks	10^{-4}
sedimentary rocks	10^{-5}

A distribution of measured (true) susceptibilities for various rock types is shown in the table. Basic refers to those rocks high in iron and magnesium silicates and magnetite, extrusive means formed by cooling after extruding onto the land surface or seafloor. The data in each category are based on at least 45 samples.

Measured susceptibilities for rock types				
rock type	% of samples with magnetic susceptibility (in 10^{-6} electromagnetic units per cubic centimetre)			
	less than 100	100–1,000	1,000–4,000	greater than 4,000
Basic extrusive (e.g., basalt)	5	29	47	19
Basic intrusive (e.g., gabbro)	24	27	28	21
Granite	60	23	16	1
Metamorphic (gneiss, schist, slate)	71	22	7	0
Sedimentary	73	19	4	4

The table lists representative values for the magnetic properties J_n (natural remanent magnetization), k (susceptibility), and ratio Q_n. Natural remanent magnetization is some combination of remanences; typically TRM in an igneous rock, perhaps DRM or CRM or both in a sedimentary rock, and all with an additional VRM. The ratio Q_n is typically higher for rocks with a strong, stable remanence—*e.g.*, magnetite-rich and fine-grained extrusive rocks such as seafloor basalts.

Typical Magnetic Properties of Rocks

rock	J_n natural remanent magnetization (10^{-5}electromagnetic units per cubic centimetre)	k magnetic susceptibility (10^{-5}electro-magnetic units per cubic centimetre)	ratio* $Q_n=J_n/k \cdot H_{ex}$
Igneous			
granite	10–80	50–400	0.3–1
diabase	190–400	100–230	2–3.5
basalt	200–1,000	100–700	5–10
seafloor basalt (1–6 metres)	500–800	30–60	25–45
typical (average)	10–4,000	5–500	1–40
Sedimentary			
red sediments	0.2–2	0.04–6	2–4
sandstone		1–40	
shale		1–50	
limestone		0.5–20	
typical (average)	0.1–10	0.3–30	0.02–10
Ores			
magnetite ore	300,000–1,000,000	30,000–100,000	~10–50
hematite ore		10–70	

*For external magnetic field (Hex) = 0.5 oersted, the cgs electromagnetic unit of magnetic field intensity.

Ore Mineral

Ore mineral is defined as a mineral which contains a metallic element in a quantity that can be exploited and extracted for use at an economical cost. Clays are rich in aluminum but since metal cannot be extracted from ordinary clays at an economical cost, these are not included among the ores of aluminum.

Importance

Ore minerals are generally oxides, sulfides, and silicates. In addition, they may be "native" metals (such as copper) that are not commonly concentrated in the Earth's crust, or "noble" metals (not usually forming compounds) such as gold. The ores must be processed to extract the metals of interest from the deposit.

- Argentite: silver sulfide (Ag_2S).
- Barite: barium sulfate ($BaSO_4$).
- Bauxite: mixture of aluminum oxides and hydroxides, used for producing aluminum.

- Beryl: beryllium aluminum cyclosilicate ($Be_3Al_2(SiO_3)_6$).
- Bornite: a sulfide of copper and iron (Cu_5FeS_4).
- Cassiterite: an oxide of tin (SnO_2).
- Chalcocite: copper(I) sulfide (Cu_2S), for production of copper.
- Chalcopyrite (or "peacock pyrite"): copper iron sulfide ($CuFeS_2$).
- Chromite: iron magnesium chromium oxide ($(Fe,Mg)Cr_2O_4$), for production of chromium.
- Cinnabar: red mercury(II) sulfide (HgS), for production of mercury.
- Cobaltite: cobalt, iron, arsenic sulfide ((Co,Fe)AsS).
- Columbite: Tantalite or Coltan: oxide mixture containing iron, manganese, niobium, and tantalum ($(Fe,Mn)(Nb,Ta)_2O_6$).
- Galena: lead sulfide (PbS).
- Gold: The metal gold (Au) is typically associated with quartz or is found as placer deposits.
- Hematite: iron(III) oxide ($Fe2_2O3_3$).
- Ilmenite: a crystalline form of iron titanium oxide ($FeTiO_3$).
- Magnetite: iron(II,III) oxide (Fe_3O_4), a ferrimagnetic mineral.
- Molybdenite: molybdenum disulfide (MoS_2).
- Pentlandite: a sulfide of iron and nickel ($(Fe,Ni)_9S_8$).
- Pyrolusite: manganese dioxide (MnO_2).
- Scheelite: calcium tungstate ($CaWO_4$).
- Sphalerite: zinc sulfide (ZnS), with variable amounts of iron.
- Uraninite (pitchblende): mainly uranium dioxide (UO_2), used for production of metallic uranium.
- Wolframite: a tungstate of iron and manganese ($(Fe,Mn)WO_4$).

Ore Deposit

Natural concentration of an ore mineral in a massive rock body is defined as an ore deposit. It may be of any shape and form but its size is of real importance. A small patch or layer of bauxite, say one meter thick and a kilometer long will not make an ore deposit. The deposit has to be of considerable size and volume, often in terms of millions of tons, to qualify as an ore deposit. Hence the definition of an ore deposit is also size dependent.

Gangue Minerals

A few non-metallic minerals are often found associated with an. These have to be separated from the ore before the same is processed for extraction of the metal. These associated minerals are called gangue minerals and are worthless. Quartz (SiO_2) is a common gangue mineral associated with many metallics.

Origin of Ore Deposits

Mineral deposits may occur independently forming layers and bodies of different shapes extending for many kilometers in areal extent and upto many meters in depth. Sedimentary deposits of iron oxides and hydrous aluminum oxide are of this type. In other cases, the ore deposits (or more precisely s) may be present in an enclosing rock, called the country rock or the host rock.

In such cases, the host rock itself is removed in the first stage of mining. The s are then separated at the second stage of exploitation.

On the basis of their time of formation relative to the host rock, the ore deposits are divided into two groups:

1. Syngenetic and
2. Epigenetic.

(1) Syngenetic deposits are such deposits that have formed simultaneously with the enclosing rock. These are either of igneous or of sedimentary types.

(2) Epigenetic deposits are those deposits that were formed subsequent to the formation of the host rock. Epigenetic deposits of igneous, sedimentary and metamorphic types are known.

In another common method of classification, s are divided into hypogene (also termed primary) deposits and supergene (also known as secondary) deposits. The secondary deposits are formed as alteration products of primary deposits through weathering and related processes.

An ore deposit may be either of igneous or of sedimentary origin. Some deposits are related to metamorphic processes and are as such classified as deposits of metamorphic origin. Within these three broad modes of origin, a mineral deposit of economic value might have originated due to set of simple or complex processes.

Types of Ore Deposits

Igneous Deposits

These deposits are formed by crystallisation of magmatic melts in different environments.

Principal types are

a. Magmatic Deposits and
b. Hydrothermal Deposits.

Magmatic Deposits

During the crystallisation of a magmatic melt, many minerals of economic value also get formed depending on the original composition of the melt. These minerals may get concentrated towards the bottom or margins or other places within the cooling body of the magma because of some factors operating at the time of crystallisation. These concentrations are sometimes of such size and nature that their economic exploitation is feasible. Magmatic deposits form major source of some important metals as gold, chromium, titanium and uranium.

Concentration of the economic magmatic minerals may take place during the earlier or the later stages of the process of crystallisation. Accordingly, these are often distinguished as early or late magmatic crystallizations. Their stage of formation is easily revealed by their relationship with the enclosing rocks.

Magmatic ore deposits commonly occur in the following forms:

1. Segregations,
2. Disseminations
3. Injections.

Segregations

In magmatic segregations, economic minerals get concentrated more or less at the place of their formation such as along bottom or margins of an igneous body. In the earlier stages of crystallisation, when most of the magma is still in a molten state, minerals that happen to form sink to and settle at the bottom due to gravity.

This is naturally referred as gravitative settling. Similarly, towards the later stage, when most of magma has crystallized into solid rock, some left over melt containing economic minerals crystallises along the margins of the solidified magma. This process is referred as residual magmatic segregation.

Chromium and titanium ores are examples of early and late segregations respectively.

Disseminations

In magmatic disseminations, the economic minerals are dispersed at random in a cooled igneous body. In some cases the dispersion is within a limited space whereas in other cases it is over a large volume when concentration is economically not exploitable. When the mineral of economic value is dispersed only in a limited space in the body of igneous mass and the trend is common over large area, whole of the rock mass is mined for exploitation and extraction of the valuable mineral.

Diamond disseminations are a typical example. This is its mode of occurrence in the famous diamond pipes of South Africa where the mineral occurs as dispersed crystals in ultrabasic rocks called Kimberlites. In India, the Panna diamond mines of Madhya Pradesh are also similar in nature to Kimberlites of South Africa. The main diamondiferous pipe is of volcanic origin and has a width of more than 350 metres, made up primarily of serpentine rock.

Magmatic Injections

These are thin, lenticular or slightly tabular bodies of magmatic minerals that appear as having been intruded or injected into the host rocks. When magma rises up with sufficient hydrostatic force, it may intrude into available fissures and fractures in the surrounding rocks where eventually it may cool down. Injection deposits may result both as early and late stages of crystallization.

Pegmatite Deposits

These are sometimes classified as a distinct group of magmatic deposits formed towards the end of crystallization process and as such necessarily occurring close to the roofs of magmatic masses. The pegmatites may be either of simple type or of complex type.

In the simple pegmatites, the rock is made up primarily of potash felspars (e.g. orthoclase) and quartz whereas in the complex pegmatites, different zones can be distinguished in the rock which is indication of differentiation having taken place as per a definite pattern.

In their form, pegmatites occur as dikes, veins, lenses and nests of variable sizes and dimensions. Some pegmatites are known for the exceptionally large-sized crystals they have yielded. Crystals of mineral beryl, a typically pegmatite mineral, as big as 6 m long, 1.3 m in diameter and 18 tons in weight have been obtained at Maine, at Abany in U.S.A.

These rocks (pegmatites) are often associated with granites (granite-pegmatite). Pegmatites associated with basic and ultra-basic rocks are also known. Pegmatites of a recrystallised type called Metasomatic Replacement Pegmatites are also of common occurrence.

Pegmatites are source of mica, quartz and many strategic minerals such as beryl, lithium minerals, rare earths and numerous gems.

Hydrothermal Deposits

This is another distinct group of economic minerals, which has been formed from cooling of gaseous and liquid solutions in cavities, fissures or pore spaces of the rocks, that is, wherever these solutions find a place to enter. The solvent in such cases is very often, superheated steam, emanating from magmas towards the final stages of crystallization.

The three essential conditions for the formation of hydrothermal deposits are – highly active and enriched fluids, suitable pathways (cracks, fissures etc.) for their migration through the rocks, and suitable physico-chemical environment for their deposition to take place.

As regards superheated steam, it is an essential component of many magmatic melts and remains dispersing around the magma body till it cools down to a solid mass. The steam is capable of containing within it, in dissolved form, metallic elements like gold, copper, tungsten, molybdenum and to some extent silver, lead and zinc etc. These elements are crystallized out as the enriched solutions pass upwards and suffer cooling and loss of pressure.

The fissures or fractures through which these fluids travel may be of primary or secondary character. Bedding planes, porosity and jointing structures are examples of primary spaces. Faults, solution cavities, fissures and jointing caused by stresses in the host rocks after their formation are examples of common spaces of secondary origin.

The question of describing the physico-chemical environment responsible for deposition of economic minerals from the hydrothermal solution remains yet unanswered. Many hypotheses have been put forward. The problem is of a very complex nature and solution may lie in one or more of the following possibilities.

Deposition may take place due to:

1. A change in temperature or pressure (or both, of the ascending solutions);
2. Ensuring chemical reactions between various components present within these liquid-gas solutions;
3. Chemical reactions between the components of the solutions and those of the country rock.

Such factors as change in the pH of the solutions, coagulation of colloids (a form in which elements might be in transport) and effects of natural electric fields have also been mentioned as (possibly) being responsible for deposition of mineral deposits from the hydro-thermal solutions.

Hydrothermal deposits occur in a variety of shapes and sizes.

The most common broad based forms are:

1. Veins and
2. Cavity fillings.

Veins

These are narrow, elongated or tabular bodies of economic minerals occurring within a host rock of entirely different composition and origin.

These are further distinguished into following types

(1) Fissure-Veins

These may be defined as mineral bodies of elongated or tabular shape deposited in pre-existing fissures. The original fissures may be parallel or intersecting, radial or fan-shaped in pattern and that is also the shape of the resulting ore bodies.

(2) Ladder-Veins

These are commonly found in igneous rocks such as dikes and similar bodies and consist of transverse, roughly regularly spaced fractures that are filled with deposits of economic minerals. These veins give a ladder-like impression in vertical section, hence the name

(3) Gash-Veins

These are narrow, sloping and thick-bodies deposits of minerals deposited in solution-fissures or cracks of the host rock which is generally a carbonate rock like limestone.

(4) Stock Works

The term is used for hydrothermal deposits that occur in veins or fissures of exceptionally small size, but in good number, within a limited space, traversing the body of rock profusely. Mining in such cases involves removal of the entire rock body containing the stock work.

Cavity Fillings

They are such ore bodies that get deposited from the hydrothermal fluids in well- defined open spaces (rather than along fractures or cracks) available in the host rock. A small or big cavity (caused due to any reason) occurring in a rock when got filled with a mineral deposit would qualify as a cavity filling. It may be regular or irregular in outline. The resulting deposit from cavity fillings may be quite massive.

Another interesting example of filling is provided by the so-called saddle-reefs. These are mineralized cavities associated with the folded rocks comprising beds of competent and incompetent nature. In such cases hydrothermal solutions find pathways in the weak zones along axial regions and deposit their minerals in those zones.

Metasomatic Deposits

These deposits are formed essentially by an action of chemically active fluids (gases and liquids) involving simultaneous solution of pre-existing component(s) from a rock and replacement of the same by new mineral(s).

The basic difference between the pure hydrothermal (magmatic) deposits and metasomatic deposits lies in the process of replacement of some original component by a new component in a solid state change.

The metasomatic (replacement) deposits of economic minerals are important source of many metals and non-metals.

The process of metasomatic replacement may be quite simple or complex in nature. Both liquid and gaseous solutions may be individually or jointly responsible for the change to take place. Liquid solutions are, however, credited with a major role in the formation of the ore deposits.

Temperature and pressure are not binding factors for replacement to occur; replacement deposits formed at surface temperature and pressure are well known. But the fact remains that at elevated temperatures, the process of replacement is accelerated in many cases.

Further, though every rock may be susceptible to replacement process, it has been observed to occur commonly in rocks of suitable composition only which belong to the carbonate group, i.e. limestones and dolomites. Igneous rocks have also been found altered by replacement (metasomatic) process. Sandstones, quartzites and argillaceous rocks like clays and shales are least susceptible.

The replacement deposit may occur in the form of veins, lodes or zones. These are commonly termed as sheet lodes or sheet zones.

The metasomatic replacement deposits may be differentiated from the magmatic and hydrothermal deposits by virtue of replacement evidences which include:

- Preservation of original rock structure;
- Presence of small sized fragments of the original rocks (still unaltered);
- Typical outlines of the ore bodies.

Metasomatic replacement deposits have been the source:

- Metals: like beryllium, caesium, lithium, niobium, rubidium, tantalum, thorium, tin, tungsten and uranium;
- Non-metals: quartz and fluorite;
- Gemstone: amethyst, aquamarine, garnet and topaz.

Sedimentary Deposits

Some very important deposits of ores of iron, copper, gold, phosphates and coal are of sedimentary origin.

Their accumulation takes place due to one or more of the following processes:

(a) Weathering

Atmospheric gases and water vapour are incessantly reacting with surface rocks of suitable composition whereby they get decomposed and altered. The process of change is called weathering of rocks. The altered, disintegrated rock product is called mantle of waste.

This mantle is the source material for development of economic mineral deposit which can take place by either of the following two ways:

First

Most of the useless (economically speaking) materials from the loose mantle is leached away or removed by subsurface waters leaving behind residual material that has economic importance. Such concentrations form Residual Deposits. Bauxite, the well-known ore of aluminium, is a typical residual accumulation in most cases.

Ores of iron like limonite, goethite and hematite and of manganese like psilomelane and pyrolusite are other examples of residual deposits. Similarly, deposit of silicate nickel ore of residual origin is also known. The kyanite deposits of Singhbhum (India) occur in the form of residual boulders.

Second

The soluble components of mantle of waste are taken into solution by percolating waters and re-deposited at lower levels during the process of infiltration. These deposits, when of economic value and size form Infiltration type of deposits. Among important deposits of this type may be mentioned ores of uranium, silicate-carbonate group ores of iron, some types of bauxite and also hydroxide of iron, native sulphur and hydrous calcium sulphate (gypsum).

(b) Sedimentary Genesis

Such deposits originate due to set of natural processes in almost an orderly fashion which involves:

- Physical and chemical disintegration of rocks by weathering and erosion into sediments.

- Transport of these sediments by wind, water and ice to the suitable environment of deposition.
- Deposition, followed by diagenesis of these sediments into sedimentary formations.

The deposits of pure sedimentary origin occur very commonly in the form of layered formations of extensive areal extent and considerable depth. In economic value they range from non-metallic deposits like sandstones and limestones (useful as building stones), clays, salts of sodium, potassium and magnesium to those of important metals like iron, manganese, aluminium, copper, uranium and vanadium etc. Coals also form a distinct group of sedimentary deposits in which biochemical changes have also taken place during the process of deposition and diagenesis.

(c) Placer Deposits

These may be broadly described as a special group of weathering deposits that have been accumulated in varied environments by virtue of some particular qualities of the minerals involved. Among such qualities a high density and chemical stability are most important. Similarly good strength and hardness against abrasive actions are also essential qualities.

These economic minerals or substances once released from the original rock masses due to processes of physical and chemical weathering are transported to places of deposition by such factors as gravity, wind, water and ice.

Accordingly, placer deposits are often distinguished on the basis of the agency involved in their transport and deposition, such as:

- Deluvial placers, when gravity is the agency involved; such placers would be found along the foot hills and talus slopes;
- Aeolian placers, when sorting and transport is carried out by the agency of wind;
- Alluvial placers, where running water such as of streams and rivers is involved in their transport and deposition;
- Beach placers are deposits of economic value accumulated by waves along shores of water bodies like seas, lakes and oceans.

Placers deposits found resting on the rocks from which they are generally derived are often called eluvial placers.

The formation of placer deposits is primarily due to sorting power of the transporting medium and the inherent physical properties of ore or other minerals. Heavier, stronger and stable minerals have better chances of accumulation and preservation as deposits whereas lighter and chemically unstable minerals get further disintegrated and dispersed.

Metamorphic Deposits

Metamorphism is a natural process of change brought out in rocks of all types subjected to changed conditions of temperature, pressure and chemically active fluids. In this process, pre-existing rocks and minerals of economic value may undergo metamorphic changes. Such deposits may be simply called as metamorphosed igneous or sedimentary deposits. In a truly metamorphic deposit, however, new minerals or rocks of some economic value are formed as a result of metamorphic change.

Genesis of Common Ores

Specific ores are organized here according to the metal commodities.

Iron

Iron ores are overwhelmingly derived from ancient sediments known as *banded iron formations* (BIFs). These sediments are composed of iron oxide minerals deposited on the seafloor. Particular environmental conditions were needed to transport enough iron in seawater to form these deposits, such as acidic and oxygen-poor atmospheres in the Proterozoic Era.

In addition, weathering during the Tertiary or Eocene periods converted the usual magnetite minerals into hematite, which is more easily processed. Some iron deposits in the Pilbara of West Australia are *placer deposits*, formed by the accumulation of hematite gravels called *pisolites*. They are less expensive to mine.

Lead, Zinc, Silver

Lead-zinc deposits are generally accompanied by silver, hosted within the mineral galena (lead sulfide) or sphalerite (zinc sulfide).

Lead and zinc deposits are formed by the discharge of deep sedimentary brine onto the seafloor (termed SEDEX deposits), or by the replacement of limestone in skarn deposits, or by subvolcanic intrusions of granite. The vast majority of lead and zinc deposits are Proterozoic in age.

- SEDEX type deposits: Examples are the immense Broken Hill, Century Zinc, Lady Loretta, and Mt. Isa deposits in Australia; the Sullivan, Red Dog, and Jason deposits of North America; and the Hindustan zinc belt in India.
- Limestone replacement-type deposits: They are exemplified by the Mississippi Valley-type (MVT) deposit. Some of these occur by replacement and degradation of hydrocarbons, which are thought important for transporting lead.
- Subvolcanic intrusion-type of deposits: Renowned for high silver grades, they are typified by deposits in Argentina, Bolivia, and Peru. These deposits are essentially Cenozoic in age and are known as the Andean silver belt. The most recent example is San Cristobal, with 450 million ounces of silver. These deposits were formed by the discharge of fluids bearing incompatible elements from the cooling granite mass, and have low lead grades but exceptional silver enrichment.

Gold

Gold deposits are formed through a very wide variety of geological processes. The underlying mechanism is plate tectonics.

They are classified as (a) primary deposits, (b) alluvial or placer deposits, and (c) residual or laterite deposits. A deposit may contain a mixture of all three types of ore.

- The majority of primary gold deposits fall into two main categories: lode gold deposits and intrusion-related deposits.

    1. Lode gold deposits: They consist chiefly of quartz veins, also known as lodes or *reefs*, which contain either native gold or gold sulfides and tellurides. Lode gold deposits are usually hosted in basalt or in sediments known as turbidite, although when in faults, they may occupy intrusive igenous rocks such as granite. Lode-gold deposits are intimately associated with orogeny (mountain-forming processes) and other plate collision events in geologic history. Most lode gold deposits *sourced* from metamorphic rocks because it is thought that the majority are formed by dehydration of basalt during metamorphism. The gold is transported up faults by hydrothermal waters and deposited when the water cools too much to retain gold in solution.

    2. Intrusion-related gold deposits: Generally hosted in granites and porphyry, this gold usually contains copper and is often associated with tin and tungsten. Intrusion-related gold deposits rely on gold existing in the fluids associated with magma, and the inevitable discharge of these hydrothermal fluids into the wall-rocks. Skarn deposits are another manifestation of intrusive-related deposits.

- Placer deposits are secondary deposits, derived from pre-existing gold deposits. They are formed by alluvial processes in rivers and streams and on beaches. Placer gold deposits form by gravity, when the density of gold causes it to sink into trap sites in the river bed, or when water velocity drops, such as at bends in rivers and behind boulders. Placer deposits are often found in sedimentary rocks and can be billions of years old, such as the Witwatersrand deposits in South Africa. Sedimentary placer deposits are known as 'leads' or 'deep leads'.

- Laterite gold deposits are formed from pre-existing gold deposits (including some placer deposits) during prolonged weathering of the bedrock. Gold is deposited within iron oxides in the weathered rock or regolith, and may be further enriched by erosion. Some laterite deposits are formed by wind erosion of the bedrock, leaving a residue of native gold metal at the surface.

Platinum and Palladium

Platinum and palladium are precious metals generally found in ultramafic rocks (igneous rocks rich in minerals of magnesium and iron). The source of platinum and palladium deposits is ultramafic rocks that have enough sulfur to form a sulfide mineral in molten magma. The sulfide mineral gains platinum by mixing with the bulk of the magma because platinum has an affinity for sulfur and is concentrated in sulfides. Platinum may also occur in association with chromite, either in the chromite mineral itself or in sulfides associated with it. Platinum is often associated with nickel, copper, chromium, and cobalt deposits.

Nickel

Nickel deposits are generally found in two forms: sulfide and laterite.

- Sulfide-type nickel deposits are formed in essentially the same manner as platinum deposits. Nickel has an affinity for sulfur, so an ultramafic or mafic rock that has a sulfide phase in the magma may form nickel deposits. The best nickel deposits are formed where sulfide accumulates, much like in a placer gold deposit, in the base of lava tubes or volcanic flows—especially komatiite lavas.
- Nickel laterite deposits are formed by a process essentially similar to the formation of gold laterite deposits, except that ultramafic or mafic rocks are required. Generally, nickel laterites require large, olivine-bearing ultramafic intrusions. Minerals formed in laterite nickel deposits include gibbsite.

Copper

Copper is found in association with many other metals and deposit styles, including deposits of gold, lead, zinc, and nickel. Commonly, copper is either formed within sedimentary rocks or associated with igneous rocks.

The world's major copper deposits are formed within the granitic porphyry copper style. The source of copper is generally thought to be the Earth's lower crust or mantle, where the granite melt forms. The copper is enriched by processes during crystallization of the granite and forms as chalcopyrite, a sulfide mineral, is carried up with the granite. Granites sometimes move to the suface with volcaniceruptions, and copper mineralization occurs during this phase, when the granite and volcanic rocks cool via hydrothermal circulation.

Sedimentary copper forms within ocean basins in sedimentary rocks. Generally, this occurs when brines from deeply buried sediments discharge into the deep sea, precipitating copper (and often lead and zinc) sulfides directly onto the seafloor. This is then buried by further sediment.

Uranium

Uranium deposits are usually derived from radioactive granites, where certain minerals such as monazite are leached during hydrothermal activity, or during circulation of groundwater. The uranium is brought into solution by acidic conditions and is deposited when this acidity is neutralized. Generally, this occurs in certain carbon-bearing sediments, in what is called an "unconformity" in sedimentary strata. The majority of the world's nuclear power is sourced from uranium in such deposits.

Uranium is also found in nearly all coal, at several parts per million, and in all granites. Radon is a common problem during mining of uranium, as it is a radioactive gas.

Uranium is also found associated with certain igneous rocks, such as granite and porphyry. The Olympic Dam deposit in Australia is an example of this type of uranium deposit. It contains 70 percent of Australia's share of 40 percent of the global, low-cost, recoverable uranium inventory.

Titanium

Titanium ore is formed as placer deposits or within ultramafic layered intrusions. In the latter case, titanium takes the form of layers of ilmenite, a titanium oxide mineral, through the process of crystallization as the intrusion cools. These layers can be considerably heavy and long, and this type of ore is known as "hard rock titanium." In addition, the ore may contain vanadiumas a second metal within the ilmenite.

Mineral Sands

Mineral sands, a type of "placer deposits," are the predominant type of titanium, zirconium, and thorium deposits. They are formed by the accumulation of heavy minerals within beach systems. The minerals that contain titanium are ilmenite and leucoxene; zirconium is contained within zircon; and thorium is generally contained within monazite. These minerals are sourced primarily from granite bedrock by erosion and transported to the seashore by rivers, where they accumulate in beach sands. On rare but important occasions, gold, tin, and platinum deposits also form in beach placer deposits.

Tin, Tungsten and Molybdenum

Tin, tungsten, and molybdenum generally form in a certain type of granite, by a mechanism similar to that for intrusion-related gold and copper. They are considered together because the process of forming these deposits is essentially the same. Minerals of these three metals are found in an important deposit formed by a process known as skarn-type mineralization. Skarn deposits are formed by the reaction of mineralized fluids from the granite reacting with wall rocks such as limestone. Skarn mineralization is also important in the formation of ores of lead, zinc, copper, and gold, and sometimes uranium as well.

Rare Earth Elements, Niobium, Tantalum and Lithium

The overwhelming majority of rare earth elements (lanthanoids), niobium, tantalum, and lithium are found within pegmatite. Ore genesis theories for these ores are wide and varied, but most involve metamorphism and igneous activity. Lithium is present as spodumene or lepidolite within pegmatite. In addition, carbonatite intrusions are an important source of these elements.

Phosphate

Immense quantities of "phosphate rock" occur in older sedimentary basins, generally formed in the Proterozoic. Phosphate deposits are thought to be sourced from the skeletons of dead sea creatures that accumulated on the seafloor. Similar to iron ore deposits and oil, particular conditions in the ocean and environment are thought to have contributed to these deposits in the geological past.

Phosphate deposits are also formed from alkaline igneous rocks such as nepheline syenites, carbonatites, and associated rock types. In this case, the phosphate is contained within magmatic apatite, monazite, or other rare-earth minerals.

Mineral Physics

Mineral physics is the materials science of geological and geo-inspired synthetic materials, with emphasis on variations of structure, properties, and behavior of minerals, melts, and fluids as a function of pressure and temperature. Our field is transitioning from assembling an immense dossier of information about these Earth materials to elucidating the dynamic response of Earth to planetary-scale driving forces.

The field of mineral physics provides ever-improving measurements of and constraints on material behavior that refine interpretation of growing suites of geophysical and geochemical observations. Groundbreaking discoveries driven by technological advances in materials synthesis, experiments, and analysis stimulate development of new hypotheses about the inner workings of Earth and other planets. As a result, mineral physics has grown hand in hand with the geosciences, responsive to the needs of the broader scientific community while simultaneously catalyzing transformative advances.

Mineral physics has proven very adaptable to technological advances and has been able to quickly and creatively take advantage of these new capabilities.

Mineral physics has a central place among solid Earth science disciplines because the well-defined, quantitative information and constraints it generates allow expression of complex, large-scale geological processes and interpretation of indirect seismic observations by means of simple chemical and physical models. Despite a century of active research, mineral physics remains a dynamic field and continues to produce many exciting discoveries, some of which change the generally accepted picture of Earth's interior.

Mineral physics is a prominent generator of both quantitative and qualitative thermodynamic information used by researchers in geodynamics, seismology, petrology, and geochemistry.

Mohs Scale of Mineral Hardness

Mohs Hardness Kit: A laboratory Mohs Hardness Scale kit containing: (1) talc; (2) gypsum; (3) calcite; (4) fluorite; (5) apatite; (6) orthoclase; (7) quartz; (8) topaz; and (9) corundum. Diamond is not included in most kits to keep the cost down. Also a diamond specimen would be so small that it would need to be mounted in a handle to be useful. Purchase a Mineral Hardness Kit.

One of the most important tests for identifying mineral specimens is the Mohs Hardness Test. This test compares the resistance of a mineral to being scratched by ten reference minerals known as the Mohs Hardness Scale. The test is useful because most specimens of a given mineral are very close to the same hardness. This makes hardness a reliable diagnostic property for most minerals.

Friedrich Mohs, a German mineralogist, developed the scale in 1812. He selected ten minerals of distinctly different hardness that ranged from a very soft mineral (talc) to a very hard mineral (diamond). With the exception of diamond, the minerals are all relatively common and easy or inexpensive to obtain.

Making Hardness Comparisons

"Hardness" is the resistance of a material to being scratched. The test is conducted by placing a sharp point of one specimen on an unmarked surface of another specimen and attempting to produce a scratch. Here are the four situations that you might observe when comparing the hardness of two specimens:

1. If Specimen A can scratch Specimen B, then Specimen A is harder than Specimen B.
2. If Specimen A does not scratch Specimen B, then Specimen B is harder than Specimen A.
3. If the two specimens are equal in hardness then they will be relatively ineffective at scratching one another. Small scratches might be produced, or it might be difficult to determine if a scratch was produced.
4. If Specimen A can be scratched by Specimen B but it cannot be scratched by Specimen C, then the hardness of Specimen A is between the hardness of Specimen B and Specimen C.

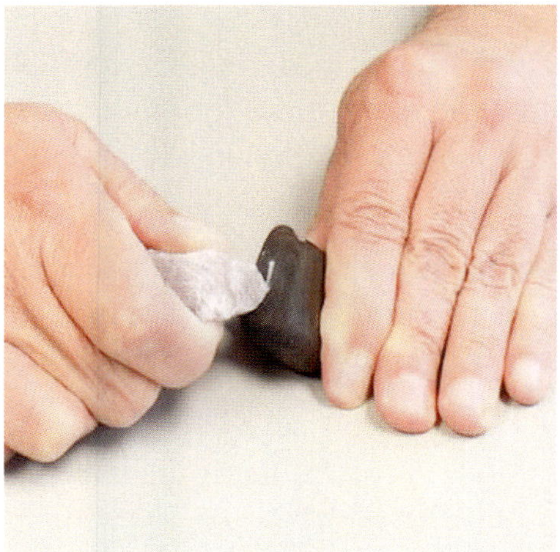

Mohs hardness test: When conducting the test, place the unknown specimen on a table top and firmly hold it in place with one hand. Then place a point of the reference specimen against a flat, unmarked surface of the unknown specimen. Press the reference specimen firmly against the unknown, and deliberately drag it across the flat surface while pressing firmly. To avoid injury, drag the known specimen away from your body and parallel to the fingers that are holding the unknown specimen.

Mohs Hardness Testing Procedure

- Begin by locating a smooth, unscratched surface for testing.
- With one hand, hold the specimen of unknown hardness firmly against a table top so that the surface to be tested is exposed and accessible. The table top supports the specimen and helps you hold it motionless for the test.
- Hold one of the standard hardness specimens in the other hand and place a point of that specimen against the selected flat surface of the unknown specimen.
- Firmly press the point of the standard specimen against the unknown specimen, and firmly drag the point of the standard specimen across the surface of the unknown specimen.
- Examine the surface of the unknown specimen. With a finger, brush away any mineral fragments or powder that was produced. Did the test produce a scratch? Be careful not to confuse mineral powder or residue with a scratch. A scratch will be a distinct groove cut in the mineral surface, not a mark on the surface that wipes away.
- Conduct the test a second time to confirm your results.

Mohs Hardness Testing Tips

- A list of minerals in order of hardness can be a handy reference. If you determine that a specimen has a hardness of Mohs 4, you can quickly get a list of potential minerals.
- Practice and experience will improve your abilities when doing this test. You will become faster and more confident.
- If the hardness of the unknown specimen is about 5 or less, you should be able to produce a scratch without much exertion. However, if the unknown specimen has a hardness of about 6 or greater, then producing a scratch will require some force. For those specimens, hold the unknown firmly against the table, place the standard specimen against it, press firmly with determination, then holding pressure slowly drag the standard specimen across the surface of the unknown.
- Don't be fooled by a soft standard specimen producing a mark on a hard unknown. That mark is like what a piece of chalk produces on a blackboard. It will wipe off without leaving a scratch. Wipe your finger across the tested surface. If a scratch was produced, there will be a visible groove. If marks wipe away then a scratch was not produced.
- Some hard materials are also very brittle. If one of your specimens is breaking or crumbling rather than scratching, you will have to be very careful while conducting the test. Testing tiny or granular specimens can be difficult.
- Some specimens contain impurities. If the results of your test are not visibly conclusive, or if the information from your test does not conform with other properties, do not hesitate to do the test again. It is possible that a small piece of quartz (or another impurity) was embedded in one of your specimens.
- Don't be wimpy. This is a very common problem. Some people casually rub one specimen back and forth against another and then look for a mark. That is not how the test is done. It

is done with a single, determined motion with the goal of cutting a scratch.

- Be careful. When you hold the unknown specimen against the table, position it so that the known specimen will not be pulled across one of your fingers.
- This test should be done on a lab table or work bench with a durable surface or a protective covering. Don't do this type of testing on fine furniture.
- Test tiny particles or grains by placing them between two pieces of an index mineral and scraping them together. If the grains are harder than the index mineral, scratches will be produced. If the grains are softer they will smear.

Hardness of Common Objects

Mohs Hardness of Common Objects	
fingernail	2 to 2.5
copper	3
nail	4
glass	5.5
knife blade	5 to 6.5
steel file	6.5
streak plate	6.5 to 7
quartz	7

Some people use a few common objects for quick hardness tests. For example, a geologist in the field might always carry a pocket knife. The knife can be used for a quick hardness test to determine if a specimen is harder or softer than Mohs 5 to 6.5.

Before using these objects as quick testing tools, it is a good idea to confirm their hardness. Some knives have harder steel than others. Test yours and then you know its hardness.

These common objects can also be useful if you don't have a set of reference minerals. We included quartz in this list because it is a ubiquitous mineral. In the field you are often no more than a few steps away from a piece of quartz.

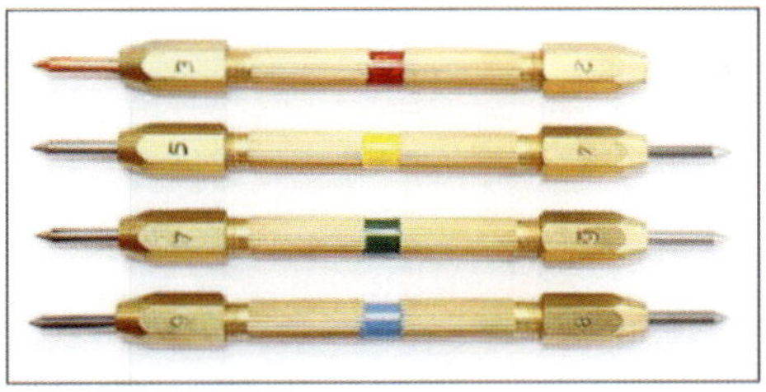

Mohs hardness picks: Hardness picks are easy to use. They have a brass stylus and an alloy «pick» that is used for hardness testing. Place the sharp point of a pick on your unknown specimen and drag it across the surface. It will either produce a scratch, slide across the surface, or leave a trace of metal. They are supplied with a hardness of 2 (a plastic point), 3 (a copper point), and 4 through 9 (carefully selected alloys). They are great for testing small specimens or for testing small grains embedded in a rock.

Hardness Picks

An alternative to using the reference minerals for testing is a set of "hardness picks." These picks have sharp metal points that you can use for very accurate testing. The picks allow much more control, and their sharp points can be used to test small mineral grains in a rock.

The sharp picks can be used easily and either produce a scratch if they are harder than the specimen being tested or leave behind a tiny streak of metal if they are softer. Examine the test site with a hand lens to see the results of your test.

We have used hardness picks and think that they do a great job. They are easier to use and more accurate than testing with specimens. They can be resharpened when they dull. The only downside is their price (about $80 per set).

Harder than Diamond, Softer than Talc

Diamond is not the hardest substance known, but the materials that are harder are much more rare. Researchers have reported that wurtzite boron nitride and lonsdaleite can be harder than diamond.

It is unlikely that you will find a mineral that is softer than talc. However, a few metals are softer. These include: cesium, rubidium, lithium, sodium, and potassium. You will probably never need to test their hardness.

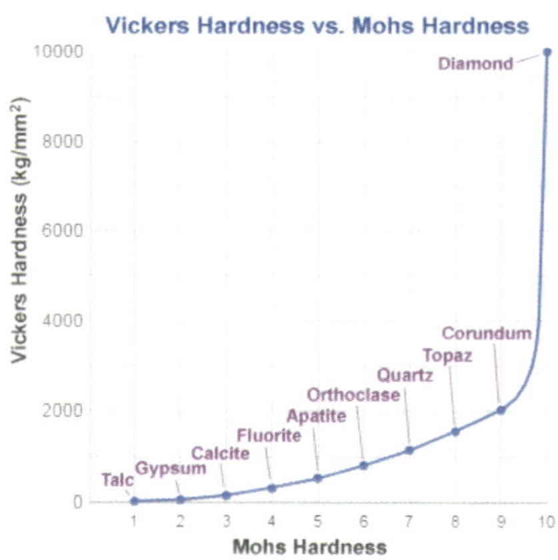

Mohs - Vickers hardness comparison: This chart compares the hardness of index minerals of the Mohs hardness scale (an integer scale) with their Vickers hardness (a continuous scale). Mohs hardness is a resistance to being scratched, while Vickers hardness is a resistance to indentation under pressure. The graph shows the great difference between the Vickers hardness of corundum and diamond - which are only one unit apart on the Mohs hardness scale.

Mohs Scale of Hardness Compared to Others

Mineral Hardness Scales		
Mineral	Mohs	Vickers (kg/mm²)
Talc	1	27

Gypsum	2	61
Calcite	3	157
Fluorite	4	315
Apatite	5	535
Orthoclase	6	817
Quartz	7	1161
Topaz	8	1567
Corundum	9	2035
Diamond	10	10000

When Friedrich Mohs developed his hardness scale in 1812, very little information about mineral hardness was available. He simply selected ten minerals that varied in hardness and arbitrarily placed them on an integer scale from 1 to 10. It was a relative scale in which a mineral of unknown hardness could be tested against a group of ten index minerals to see where it positioned on the scale.

The Mohs scale has stood the test of time and has been widely used throughout the world for over 200 years - mainly because it is easy-to-do, inexpensive and people quickly understand it. Other hardness tests have been devised but none of them are in as widespread use.

A "Mohs hardness" is a relative integer-scale comparison of "resistance to being scratched." Most other hardness scales use "resistance to indentation under a stylus to which a specific amount of pressure is applied for a specific length of time." Although these tests differ from Mohs hardness in their procedure, they are all tests of the resistance to atoms being dislodged from their positions by pressure against the surface of a mineral specimen.

One of these scales is the Vickers Hardness Scale. In the Vickers test, the size of the indentation is microscopically estimated and used to calculate a hardness value. The Vickers hardness values form a continuous scale which provides more information about the hardness of minerals when compared to the integer values of the Mohs scale. A table comparing the Mohs scale minerals to their Vickers hardness is shown here along with a graph of the data. The graph shows that in terms of Vickers hardness, the gaps between the integer values of the Mohs scale are not uniform in width. In addition the gaps between minerals of higher Mohs hardness are much broader than those between the softer minerals. In terms of Vickers hardness, diamond is enormously harder than corundum.

Hardness Variations in a Single Mineral

Although reference books and websites often list a single hardness for each mineral, many minerals have variable hardness. They have greater or lesser hardness depending upon the direction in which they are being scratched.

A well-known example of a mineral with variable hardness is kyanite. Kyanite frequently occurs in blade-shaped crystals. These crystals have a hardness of about 5 if they are tested parallel to the long axis of the crystal, and a hardness of about 7 if they are tested parallel to the short axis of a crystal. Why? These different directions encounter different bonding environments in the kyanite crystal. The bonds that resist scratching parallel to the long axis of the bladed crystal are weaker than those encountered when scratching across the width of the crystal. Intermediate hardnesses are encountered in other directions.

Another example is diamond. The people who cut diamonds have known about its variable hardness for hundreds of years. They know that parallel to the octahedral crystal faces, a diamond crystal is almost impossible to saw and very difficult to polish. The diamond can be broken in this direction by cleaving, and the best method for cutting it in this direction is with a laser. The softest and best direction to saw or polish a diamond crystal is parallel to its cubic crystal faces. This information is critical knowledge for the craftsmen who plan the design of a faceted diamond. Understanding it and working with it saves time, saves money and creates a better product with less waste.

Weathering can also influence the hardness of a mineral specimen. Weathering changes a mineral's composition, with the weathering product usually softer than the original material. When testing the hardness or streak or other property of a mineral, the best way to test is on a freshly broken surface with expected luster that has not been exposed to weathering.

About Hardness Tests

The hardness test developed by Friedrich Mohs was the first known test to assess resistance of a material to scratching. It is a very simple but inexact comparative test. Perhaps its simplicity has enabled it to become the most widely used hardness test.

Since the Mohs Scale was developed in 1812, many different hardness tests have been invented. These include tests by Brinell, Knoop, Rockwell, Shore and Vickers. Each of these tests uses a tiny "indenter" that is applied to the material being tested with a carefully measured amount of force. Then the size or the depth of the indentation and the amount of force are used to calculate a hardness value.

Because each of these tests uses a different apparatus and different calculations, they can not be directly compared to one another. So if the Knoop hardness test was done, the number is usually reported as a "Knoop hardness." For this reason, Mohs hardness test results should also be reported as a "Mohs hardness."

Why are there so many different hardness tests? The type of test used is determined by the size, shape, and other characteristics of the specimens being tested. Although these tests are quite different from the Mohs test, there is some correlation between them.

Hardness, Toughness and Strength

When testing for hardness, remember that you are testing "the resistance to scratching." During the test, some materials might fail in other ways. They could break, deform, or crumble instead of scratching. Hard materials often break when subjected to stress. This is a lack of toughness. Other materials might deform or crumble when subjected to stress. These materials lack strength. Always keep in mind that you are testing for the resistance to being scratched. Don't be fooled by other types of failure in the specimen being tested.

Uses for Hardness Tests

The Mohs Hardness Test is almost exclusively used to determine the relative hardness of mineral specimens. This is done as part of a mineral identification procedure in the field, in a classroom, or in a laboratory when easily identified specimens are being examined or where more sophisticated tests are not available.

In industry, other hardness tests are done to determine the suitability of a material for a specific industrial process or a specific end-use application. Hardness testing is also done in manufacturing processes to confirm that hardening treatments such as annealing, tempering, work hardening, or case hardening have been done to specification.

Streak

Streak is the color of a crushed mineral's powder. The color of a mineral's powder may differ from the actual color of the mineral. This property can be useful for mineral identification.

Almost every mineral has an inherent streak color, no matter what color the actual mineral is. For example, Calcite occurs in many different colors, shapes, and varieties. But every single variety of Calcite has a white streak. A streak is useful in distinguishing two minerals with the same color but different streak. A good example is distinguishng Gold (yellow streak), and Chalcopyrite.

Most light colored, non-metallic minerals have a white or colorless streak, as do most silicates, carbonates, and most transparent minerals. The streak test is most useful for identifying dark colored minerals, especially metals.

How to use Streak as an Identification Mark

When testing for streak, the mineral must be crushed to determine the color of its powder. The color of the powder is the color of the streak. Instead of actually crushing a mineral to determine the streak, it is much simpler to swipe the mineral across a streak plate. A streak plate is an unglazed piece of porcelain, such as the underside of a ceramic tile. This is the most popular method of streak testing, since the color of the streak plate is white, the color of the mineral trace is easy to see. For minerals that are harder than the streak plate, this test cannot be used, since the mineral will remove tile material. This is rarely a problem, though, since most minerals where this test will be significant are softer than the streak plate (the streak plate has an average hardness of about 6½ on the Mohs scale).

Most mineral references don't make a distinction between a white and colorless streak, since the difference is minimal. A mineral with a white or colorless streak will not leave a visible streak on a streak plate.

How to Test using Streak

Swipe the mineral against the streak plate, but not hard enough to damage it. The color of the powder that results on the streak plate is the streak color.

Type Specimen

Type mineral specimens are reference samples for the definition of mineral species. The biological sciences have a long-established system of type specimens, but the for-mal designation of such samples is a relatively recent matter in mineralogy. Indeed, for a vast number of minerals, some first discovered in antiquity, and many others as recently as this century, there exist no type specimen at all. In many such cases, the best available information comes from designations such as "original material" in museum catalogues and on some old labels. Interpretations of labels, however, even if they are in the handwriting of the original describer, should always be treated with utmost caution.

The following kinds of type specimens are those accepted and approved by the Commission on New Minerals and Mineral Names and the Commission on Museums of the International Mineralogical Association. It is emphasized that the designation of type material of mineral specie is the responsibility of the senior author of the original description of that species. In the case of "old" species for which no types were designated by the senior author, qualified, tentative designations may be given by the curator(shaving custody of the originally studied material. It is most important in cases such as this that the curator acts with the utmost responsibility and caution. If there is any doubt that certain material represents the material originally described, no type designation should be conferred. Curatorial designations are not binding on the Commissions; they may be revised if controversy requires their review. In such cases, review will be by the Commission on Museums, IMA.

Holotype: A single specimen (designated by the author) from which all the data for the original description were obtained. Where portions of such a specimen have been sent to other museums for preservation, the author will designate each of these as "part of the holotype."

Cotype: Specimens (designated by the author) as those used to obtain quantitative data for the original description. Specimens examined only visually should not be considered cotypes.

Neotype: A specimen chosen by the author of a redefinition or re-examination of a species to represent the species when the holotype or cotypes cannot be found. It must be shown that every attempt has been made to locate the originally described material. Neotypes can also be designated when examination of all holotypes and cotypes has shown that the definitive unit-cell parameters and chemical composition cannot be experimentally deter mined. All neotypes require the approval of the Commission on New Minerals and Mineral Names of the International Mineralogical Association.

Both holotypes and cotypes are possible' and even advantageous, for a mineral species. The use of "holo" here is to indicate all of the necessary data were obtained from the holotype specimen. If the author of a new mineral description chooses to designate additional samples as cotypes, this is permissible. Such cotypes are designated only if they were used to obtain quantitative, but not necessary, data. Thus, a mineral species may be represented by a holotype and one or more cotypes and neotypes.

References

- What-is-mineralogy: wisegeek.com, Retrieved 15 April 2018
- 10-4-index-minerals-and-metamorphic-facies: physicalgeology.pressbooks.com, Retrieved 11 June 2018
- What-are-index-minerals-1440840: thoughtco.com, Retrieved 25 June 2018

- Types-of-rocks: geologypage.com, Retrieved 15 July 2018
- Mineral-ore-deposits-meaning-origin-and-types-geology-5773: geographynotes.com, Retrieved 15 July 2018
- Mineral-physics-high: compres.us, Retrieved 25 June 2018

Chapter 2

Classification of Industrial Minerals

Industrial minerals are mined for their use in industries or other commercial activities depending on their physical or chemical properties. The different types of industrial minerals such as pumice, kaolinite, mica, kyanite, dolomite, bentonite, feldspar, etc. have been discussed in great detail in this chapter.

Industrial minerals can be defined as minerals, single or group, where their physical or chemical properties, and not their metallic, energetic or gem properties are the main basis for industrial purposes. Examples of typical industrial minerals (and rocks) are calcium carbonate, dolomite, olivine, feldspar, nepheline syenite, quartz/quartzite, talc, apatite, graphite and mica. All of these are, or have been, produced in Norway.

Internationally, it is quite common to use the term non-metallics as a synonym for industrial minerals. In this meaning, the industrial minerals are classified into 1) Construction materials (e.g. crushed rock, aggregates, sand, gravel, building-stone), and 2) Non-construction materials. At NGU, mineral resource research is organised into Construction materials, Industrial minerals (non-construction) and Metallic ore deposits.

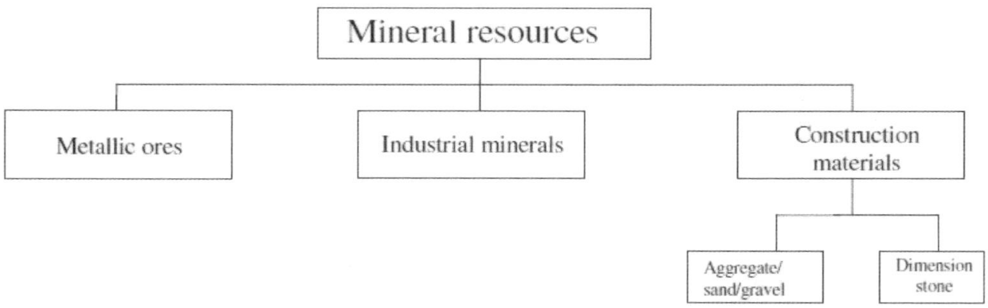

Figure: Organisation of mineral resource

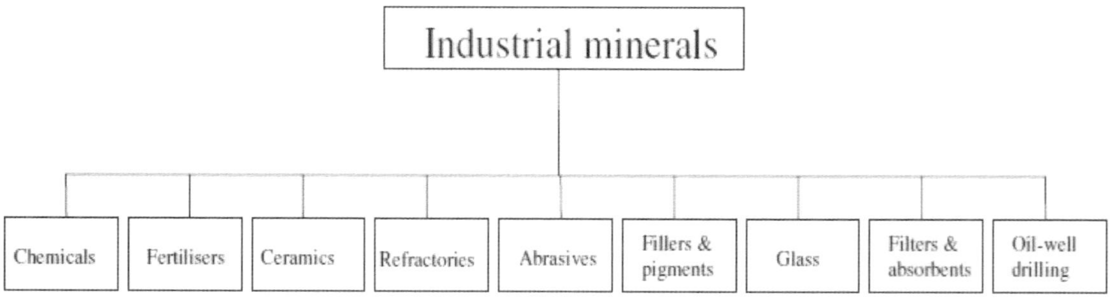

Figure: Simplified classification of industrial minerals (sensu stricto) by their enduses. There are actually many more classes

It is a common practice to classify industrial minerals according to their end-uses. In terms of the end-uses, there are a number of important though different parameters e.g. particle size distribution, chemistry, whiteness, hardness, specific gravity, chemical inertness, insulating properties, heat resistance, etc., that make the basis for different uses and the extreme diversity of products.

Most industrial minerals have a large number of different uses; for example, the mineral calcium carbonate is used as a filler in paper, paints and plastics, in cement, ceramics, fertilisers and chemicals and as a pH-regulatory material. In each of these uses there are a great number of different calcium carbonate products with different specifications. In the definition of 'industrial minerals' there is also room for metallic minerals, when properties other than their metallic properties are important; e.g. ilmenite and magnetite are both used in colour pigments and therefore fall within the definition.

The most important characteristic of industrial minerals is that they have one or more important physical or chemical properties or both. Therefore an alternative definition is proposed, which takes into account this important characteristic: "industrial minerals and rocks are Earth materials utilized because of their characteristic physical and chemical properties and not because of their metal content and which are not energy sources". This definition includes five major groups of raw materials:

- Raw materials which are utilized by the industry as minerals (e.g. talc, asbestos, micas, feldspars) or bulk rocks (e.g. bentonite, perlite, limestone) because of their special physical and chemical properties, such as mechanical strength, colour, viscosity, inertness, high melting point, cation exchange capacity etc.
- Raw materials which are source of non-metals such pyrite for sulphur, apatite for phosphorus, borates for boron and fluorite for fluorine.
- Raw materials which may be source for metals such as bauxite for Al, chromite for Cr or ilmenite for Ti, but which may also be used in certain industrial applications because of their characteristic physical and chemical properties. Hence, bauxite and chromite can be used as refractory materials and ilmenite is used as a source of TiO_2 pigment. Magnesite (source of Mg), beryllium minerals (sources of Be) and sylvite (source of K) are included in this group as well. In contrast Ni-laterites, are not included, although nickel is present in non-metallic phyllosilicate minerals (serpentine, smectite, chlorite, talc). Note that these phyllosilicates may form important industrial mineral deposits in other geological environments.
- Natural construction materials including building and ornamental stones and aggregates (granite, marble, limestone sand and gravel etc).
- Synthetic materials which may be monomineralic or may consist of several phases, such as zeolites, industrial diamonds, lime, dead burnt or caustic magnesite, smectite etc. They are synthesized from i) amorphous gels, ii) from other precursor minerals or from volcanic glass and iii) from industrial waste from the mining industry. Other materials, such as glass or cement are made of industrial minerals.

Special Features of the Industrial Rocks and Minerals

Place and Unit Value

Place value is a term used widely for industrial minerals and rocks. It expresses the influence of the geographic area in which an industrial rock or mineral is mined or quarried on its market value and reflects the importance of the value of the product in relation to its market. The unit value is the real market value of the product and depends on the development of a market for this mineral and the ability of the mining companies to add value on their end product. Added value is produced from processing which yields materials with better properties. For instance some kaolin grades for paper coating are produced by delamination, magnetic separation, flotation or leaching. Processing may sometimes yield high added value products for highly specialized markets. Typical examples are the acid activated bentonites which find application as bleaching earths for decolourising edible oils, or the organophilic bentonites, which are utilized in the formulation of nanocomposites. The production of added value industrial mineral products is feasible only when there is market available to consume these end products.

Usually industrial mineral and rock products with high place value have low unit values. For example sand and gravel or crushed aggregates are valuable only when they are consumed close to the area from which they are mined. Their value as a resource diminishes even further when they are transported away from that area, because their unit value is low and transport costs increase significantly the market value. In fact transport costs are a dominant component of the value paid by the customer. Hence aggregates in general have high place value. Nevertheless there are exceptions in this rule. The development of superquarries operating in remote areas such as those in NW Scotland, which extract hard rock aggregates, and which then are loaded to large vessels, has allowed transportation of aggregates to the Gulf coast states of the USA, the south east England and countries of north Europe. In this case the increasing demand for aggregates coupled with the lack of suitable raw materials in the consuming countries and the relatively low freight costs has allowed transportation of industrial rocks with high place value over long distances. Industrial rocks which are sold in expanded form (perlite, vermiculite) have also high place value; with a few exceptions, perlite is not transported as expanded material, but it is carried in unprocessed bulk form and undergoes expansion after transportation.

On the other hand there are many industrial rocks and minerals which have a unique physical property or combination of properties and therefore are valued by the industry. These properties impart a high unit value to the industrial minerals, which can be transported worldwide. These products are traded internationally and thus have taken a place in the international commerce. Typical examples are the kaolin deposits of Cornwall, UK (although currently there are cheaper supplies from Brazil and China), or the bentonite deposits of Milos, Greece. Transport may be facilitated by access to sea routes. For example although there are important perlite reserves in New Mexico, perlite from Greece and China is imported to the eastern and western US respectively, because sea transport is less expensive than land transport. A high unit value may be imparted also by the rarity of an industrial mineral, which is consumed by the industry. For example currently deposits of borate minerals, vermiculite, lithium minerals and rare earth element minerals are exploited only in a few countries in the world.

Significance of Physical Properties

By definition industrial minerals and rocks are utilized because of their important physical and chemical properties either as raw materials or after processing. These properties remain essentially unchanged in the end use after processing. Moreover the mode of occurrence of rocks and minerals in the field is closely related to the way in which they will be used by the industry. In some cases, the raw material cannot be utilized without processing; the latter yields new phases which are utilized because of some important physical properties. A typical example is refractories, in which the raw materials are usually fired at high temperatures to form oxides or silicates. Magnesite and dolomite are calcined at temperatures in excess of 1500° C to produce dead burnt magnesite and dolomite respectively. Similarly bauxite is fired at high temperatures to form Al_2O_3 and minerals of the kyanite group (kyanite, sillimanite, andalusite) are fired to form mullite before their use as refractories. Both dead burnt magnesite (i.e. periclase) and mullite are in the form of minerals and have high melting points.

In some industrial minerals and rocks special treatments may impart new physical properties to the raw material. For example the treatment of bentonite with inorganic acids, known as acid activation, yields materials with large number of acid centers and high sorption capacity, which can adsorb colouring compounds and decolourize and stabilize crude edible oils. This sorption capacity is not present in the raw materials. Similarly expansion of perlite or exfoliation of vermiculite yields lightweight aggregates with significant insulation properties, which are not observed in the untreated raw materials.

Competition, Substitution and Synthesis

Different industrial minerals and rocks may have similar physical and chemical properties. Since they are used by the industry for these properties it follows that there is competition for end uses and that one commodity may substitute for another in a certain industrial application. Typical examples are the substitution of kaolin by calcite or talc in paper plastic or paint fillers, the substitution of feldspar by nepheline syenite in ceramics the substitution of bentonite by palygorskite in the drilling industry etc. Substitution is affected by the intrinsic properties of the raw material, the local availability of the resource, the price and technical requirement. An example of the significance of intrinsic properties is the palygorskite drilling muds, which do not flocculate in the presence of electrolytes. Therefore, they have superior rheological properties and successfully replace bentonites in drilling using sea water. Local availability is an important factor affecting substitution of industrial minerals, because by tradition local industry is based on indigenous sources in order to suppress production costs. For example the local availability of kaolin, calcite or talc partly determines which type of filler will be used in the paper industry. The technical superiority of an imported material against local resources may encourage the development of appropriate technology to use local resources more effectively. Hence successful processing may improve the properties of local resources and increase their added value against imported raw materials.

The tendency for substitution of industrial raw materials and technological change varies with the type of industry and depends on the general economic climate, management and operational flexibility. In periods of economic recession substitution of raw materials is less likely to occur, because the industry usually undertakes minimum risks. Also substitution may be imposed by environmental reasons. In this respect zeolites have replaced phosphates in powder detergents. They are considered environmentally friendly because they act as deflocculants and water softeners and they do not contribute to eutrophication of natural waters, and thus reduce the environmental impact.

Synthetic industrial minerals often are competitors of their natural counterparts or of other raw materials. Although they have a higher production cost which increases their unit value they are preferred in certain applications in which product quality is of top priority for two reasons: a) their composition and hence properties does not vary b) their physical properties may be tailored according to the specifications of the particular industry. Often new species with superior properties can be synthesized, which do not have natural analogues. Synthetic zeolites are typical examples of synthetic minerals utilized in various industrial applications, because of their high cation exchange capacity, their catalytic properties and their action as molecular sieves due to the presence of channels of certain dimensions in their structure. They are preferred by the chemical industry over the natural zeolites, because they have a concistent and homogeneous composition. Moreover novel species with superior properties have been synthesized, which do not have natural analogues (i.e. zeolite A, zeolite ZSM-5 ect).

Although synthetic zeolites have prevailed over their natural counterparts in the chemical industry, natural zeolites are also used successfully in several applications, mainly in environmental protection, in agriculture, in aquaculture, in animal nutrition and as a building stone. Similarly, both synthetic and natural diamonds are utilized by the abrasives ndustry and natural and synthetic smectites compete in many applications. A well known synthetic trioctahedral smectite is laponite (synthetic hectorite) which has very important gelling properties. In other cases, synthetic minerals have replaced natural ones almost completely from their main application field. For example emery has been replaced by synthetic materials in most markets for abrasives. It now finds applications only in non-skid flooring and concrete hardening, which consume only small amounts.

Variable Functions in Industry

Industrial minerals and rocks and the products derived from them are interrelated in various industrial processes. The interdependence has three components. a) Use of raw materials or their products for production of another industrial mineral b) synergetic action of different industrial rocks or minerals in an industrial application and c) complementary action of different industrial minerals in industrial applications. Typical example in the former case is the production of sulphur with the Frasch process. Production of sulphur with the Frasch process requires large quantities of fresh water, which are filtered and treated by chemical compounds such as anhydrous disodium phosphate, lime and hydrated dolomitic lime sodium hydroxide, sodium carbonate sodium nitrate etc. These compounds are produced from other industrial minerals, including limestone, dolomite, sodium chloride (halite), and phosphorite. Examples from the latter case are the iron and steel industry and the drilling industries. The iron and steel industry utilizes a large variety of industrial rocks and minerals in different stages of production. Moulding sands are made of quartz, chromite, olivine or zircon grains bound by bentonite, fluxes may contain limestone, dolomite, lime or fluorspar, and refractory bricks are made of magnesite, dolomite, alumina silica or mullite. Also the drilling industry utilizes bentonite and barite to lubricate the drilling bit and stabilize the drilling fluids. These industries are totally dependent on the use of industrial minerals.

Specifications and Standards

Assessment of an industrial rock or mineral often involves comparison of the physical or chemical properties of the raw material or the processed end product with certain national or international standards which have been issued by national or international institutions (ASTM, BSI, DIN, API, OCMA etc). This is not the case in ore deposits in which only the grade of the ore or the mineral concentrate is examined by the end user. In some cases the specifications for the property in question is dictated by the end user. In this case the end user often invents a specific test which is adopted by other industrial users. For example the suitability of industrial minerals as animal litters is examined by the Westinghouse test, which has been proposed by an industrial user.

Often industrial minerals or rocks with superior physical properties, which could meet specifications and would therefore have high unit value, are utilized as low unit value raw materials. In this case the raw materials are *underutilized*. This is due to a) the limited development of industry in the country, which could utilize the raw material to gain maximum benefit or b) to the existence of alternative industrial minerals which act as substitutes of a raw material. The former is very common in carbonate rocks, both limestones and dolomites. In this case carbonates are utilized mainly as aggregates with low unit value, although they may have white colour or may be chemically pure to be used as fillers or in the chemical industry. This is because, either there are not industrial users for the raw material, or there are technological constraints (e.g. use of acid technology in the paper industry requires kaolin fillers instead of calcite fillers). The latter is common in countries with deposits of different industrial minerals, which can be used in common applications, such as dolomite and magnesite. For example Greece is a country with significant reserves of pure dolomites and good quality cryptocrystalline magnesite deposits. Mining companies do not extract dolomite for refractory use because a) the tradition is to use magnesite as long as there are sufficient reserves and b) the limited iron and steel industry in the country which would require additional refractory resources.

Economic Aspects of Industrial Minerals and Rocks

Raw materials have been utilized from ancient times and have contributed to the development of various stages of human civilization. Historical times have been divided into the Neolithic age and the bronze and iron ages according to the composition of the main tools utilized by the man. The construction of important historical monuments which belong to the world cultural heritage like the Egyptian Pyramids the Greek Acropolis or the Roman Coliseum has consumed important industrial rock resources. Also clays were utilized for the manufacture of ceramics as early as the Paleolithic age. The development of an economy and civilization based upon manufacturing industry depends on the ability of a country to have access on adequate mineral resources, either indigenous or imported via trading. In historical times after the introduction of metal processing and subsequent utilization, the role of metals was more important than that of industrial rocks and minerals in economy. This was especially the case after the industrial revolution in the mid 18th century in the UK and later in NW Europe and the USA. Since then the role of industrial minerals and rocks has gradually increased and today the value of produced industrial minerals in most developed economies exceeds that of metals.

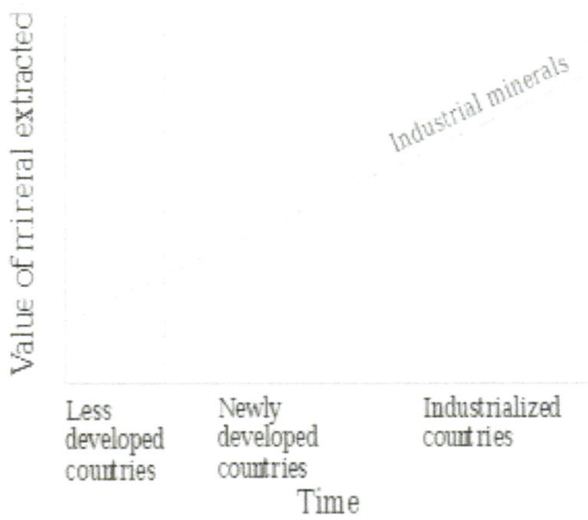

Figure: The relative value of minerals extracted in various countries

The relative importance of metallic or industrial minerals to the economy of the various countries reflects the *economic maturity* of that country. In less well developed countries metallic mineral products dominate over industrial minerals and are predominately exported to industrialized countries. With increasing development the relative importance of industrial minerals for the domestic manufacture increases. the point at which industrial minerals become more valuable than metals varies among the various countries, reflecting the industrial maturity of each country. In the UK the change was in the 19th century, in the USA in early 20th century, in Spain in the early seventies and in Australia in the late eighties. In all mature industrialized countries the value of industrial rocks and minerals by far exceeds that of metals. Figure, the value of metals is considered essentially constant through time.

In industrialized countries a majority of the industrial rocks and minerals extracted are linked to the construction industry. These materials include sand and gravel and crushed rock aggregates, (limestone and dolomite, igneous rock, sandstone), silica sand, clays and gypsum. Moreover, limestone and clays are used in the manufacture of cement, which also is consumed in the construction industry. Hence aggregate production depends entirely on the activity of the construction industry. The gradual increase of the value of industrial minerals in developed industrialized countries is related with the vast increase of the construction industry in these countries. This is related to the increasing urbanization and growing housing demands for the population, which migrates from the country to the urban areas and which is related with changes in the use of land. As a test of evidence one of the main economy sectors which have been affected by the recent economic recession is the construction industry, and this has affected adversely the production of aggregates and cement.

Talc

Talc is known for being the softest mineral on earth. It is number 1 on the Mohs hardness scale, and can be easily scratched by a fingernail.

In many industrial processes, talc is used to prevent two materials from sticking together. Talc powdering of moulds is widely used for example in tyre manufacturing or foundry applications. Certain foodstuffs or conveyor belts are also powdered with talc, to excellent effect. Similarly, chip boards are powdered to stop them from sticking together in storage. Talc is a mineral ($Mg_3Si_4O_{10}(OH)_2$ - molecular weight 379.26).

Its elementary sheet is composed of a layer of magnesium-oxygen/hydroxyl octahedra, sandwiched between two layers of tetrahedral silica. The main surfaces, known as basal surfaces, of the elementary sheet contain neither hydroxyl groups nor active ions, making talc both hydrophobic and inert.

The size of an individual talc platelet i.e. a few thousand elementary sheets, can vary from approximately 1 micron to over 100 microns depending on the conditions of formation of the deposit. It is this individual platelet size that determines a talc's lamellarity. A highly lamellar talc will have large individual platelets whereas a microcrystalline talc will have small platelets. The elementary sheets are stacked on top of one another, like flaky pastry, and because the binding forces (known as Van de Waal's forces) linking one elementary sheet to its neighbours are very weak, the platelets slide apart at the slightest touch, giving talc its characteristic softness.

Talc is odourless. It is insoluble in water and in weak acids and alkalis. Although talc has a marked affinity for certain organic chemicals, it generally has very little chemical reactivity. It is neither explosive nor flammable. Above 900° C, talc progressively loses its hydroxyl groups and above 1050° C, it re-crystallises into different forms of enstatite (anhydrous magnesium silicate). Talc's melting point is at 1500° C.

Talcs differ according to their mineralogical composition, i.e. the type and proportion of associated minerals present. The most common mineral found with talc is chlorite, which is structurally and chemically very similar. Dolomite and magnesite are also often present.

Organophilic Properties

Talc shows an affinity for certain organic chemicals. It is therefore organophilic. In the case of talc's affinity for polypropylene, a likely explanation is that the position of the oxygen atoms on the surface of the talc platelet corresponds to the carbon bonds on the surface of the polypropylene crystal.

Talc's organophilic properties are used to great benefit in a number of applications, for instance, it helps polypropylene to crystallise. In papermaking, the talc particles attract the undesirable resin droplets, i.e. organic chemicals, in the pulp onto their surface. This is known as pitch control. In body powder applications, perfume, also organic, is adsorbed onto the talc surface and retained.

Wollastonite

Wollastonite is a white to yellowish-brown naturally occurring mineral. It has good strength, firing characteristics and ability to reduce crazing. It is comprised chemically of calcium, silicon and oxygen. It's molecular formula can be expressed as $CaSiO_3$ or as $CaO.SiO_2$. It is commonly referred to as calcium metasilicate according to the now obsolete classification of silicates as oxyacids of silicon, a salt of hypothetical metasilicic acid: H_2SiO_3. It has a theoretical composition of 48.28% CaO and 51.72% SiO_2.

Wollastonite has industrial importance worldwide. It is used in many industries, mostly by tile factories which have incorporated it into the manufacturing of ceramic to improve many performance parameters, and this is due to its fluxing properties, freedom from volatile constituents, whiteness, and acicular particle shape.

Chemical Properties

Theoretical composition: CaO= 48.3% and SiO_2= 51.7%. The typical is: CaO= 47.0% and SiO_2= 50.0%.

Physical Properties

Various granulations are available. Primarily 10 X 100 mesh, 200 mesh X down and 325 mesh X down.

Typical Applications

As a source of silica and alkalis in the production of ceramics. Used to produce wall tiles, semi-vitreous bodies, ceramic insulators and glazes, thermoplastics, adhesives, and ceramic-bonded abrasives.

Wollastonite has been found useful as a soil conditioner and an additive to plant fertiliser for its reactivity in acid conditions. As opposed to lime or calcite, it offers long-term protection against the acidification of soils. In some countries, wollastonite has also found use as an inert carrier for pesticides and insecticides.

Pumice

The word pumice is derived from the Latin word pumex, meaning foam.

Pumice is a textural rock formed from volcanic eruptions. Deep underground, molten rock incorporates water and other gases, and when the magma erupts from a vent, the gases flash off, leaving behind a frothy, vesicle-riven structure that quickly cools, solidifying the foamy structure. The magma has now transformed into an amorphous aluminum silicate, or pumice.

Pumice is found in pyroclastic flows or accumulated into drifts, piles and banks (*deposits*, in mining terms) by wind or wave action.

If the newly formed pumice falls on or erupts beneath a body of water, wave action will deposit the floating mass on or near the shore as it saturates and sinks. This centuries-long process results in a very pure pumice—the heavy minerals dropping to the sea or lake floor, while the pure pumice is washed to the shoreline. This is the case with the pumice deposit that is sourced by Hess Pumice in Southeast Idaho. The Hess pumice deposit sits on the shore of an ancient inland lake known as Lake Bonneville. The current Great Salt Lake is all that is left of this once-vast freshwater lake. The vent that supplied the pumice in the Hess deposit is only about a mile away.

Mining Pumice

The various commercial pumice producers use their own variants on the process, but generally, a pumice deposit is surface-mined, meaning the overburden of soil covering the pumice is pushed aside and stockpiled for later reclamation. The pumice is then scraped up using bulldozers and pushed to the crusher for preliminary processing and stockpiling for loading and hauling to the plant for processing. When weather permits, the mining goes on year-round, otherwise, sufficient quantities are stockpiled for refining during winter months.

Pumice is then typically dried to the necessary moisture content ideal for refining before it is lifted to the top of the plant structure and via gravity, begins its journey down through the crushers, screens, and separators until it is finally bagged in sacks or bulk bags or loaded in bulk container cars or blown into pneumatic transport trailers or pneumatic rail cars.

Industrial Applications for Pumice

Pumice is an ideal aggregate for lightweight concrete building blocks.

Pumice is also used as an abrasive, especially in milder applications such as television glass, computer circuit boards, softening and aging textiles (like denim), exfoliating skin. Pumice is extremely 'friable,' which means that even when crushed to a very fine power, it does not lose its sharp edges and ability to abrade and polish.

Pumice's vesicular makeup (matrix of air-filled chambers) makes it ideal for blast mitigation, as the pourous chambers will absorb blast energy as the vesicles are crushed sequentially. When pumice is used in bomb encasements, it can help alleviate the damage from accidental blasts.

Pumice serves as a filtration media in large culinary water systems around the world. Pumice aggregate and pumice pozzolan are used in concrete applications world-wide. Pumice is the critical lightweight aggregate that underpins the run-away success of strong, lightweight manufactured stone veneer and other lightweight concrete products. Pumice aggregate increases the concrete's R-value by a factor of 4, giving pumice concrete construction a huge advantage in energy savings.

The horticulture industry uses pumice to aerate soils and add to a dense soil's ability to retain moisture. Pumice can also serve as the growing medium for plants in non-soil, hydroponic applications.

Pumice is an excellent filler. It is non-crystalline by nature, non-toxic and non-hazardous. The paint industry, for example, relies on ultra fine pumice as a functional filler that gives paint a lower density as well as burnish and stain resistance. Coarser grades of pumice provide excellent anti-skid properties. The rubber and plastics industry also uses pumice as a functional filler.

The construction industry uses pumice pozzolan enhanced concrete to add durability, longevity and strength to critical concrete applications—roads, bridges, precast concrete products, GFRCs, well cements, cementitious grouts, lightweight tilt-up panels, and decorative concrete.

Chemical Makeup of Pumice

The chemical properties of pumice will vary from deposit to deposit, but essentially, pumice is primarily Silicon Dioxide (Amorphous Aluminum Silicate), some Aluminum Oxide, and trace amounts of other oxides.

Kaolinite

Kaolinite is a clay mineral, with a soft consistency and earthy texture. It is easily broken and can be molded or shaped, especially when wet. Kaolinite is a lackluster and uninteresting mineral on its own, but it occasionally forms interesting pseudomorphs, especially after feldspars. It is also a common accessory to other minerals, including gem crystals in decomposing feldspar pegmatites.

Uses

The main use of the mineral kaolinite (about 50% of the time) is the production of paper; its use ensures the gloss on some grades of coated paper.

Kaolin is also known for its capabilities to induce and accelerate blood clotting. In April 2008 the US Naval Medical Research Institute announced the successful use of a kaolinite-derived aluminosilicate infusion in traditional gauze, known commercially as QuikClot Combat Gauze, which is still the hemostat of choice for all branches of the US military.

Kaolin is used (or was used in the past):

- In ceramics (it is the main component of porcelain)
- In toothpaste
- As a light-diffusing material in white incandescent light bulbs
- In cosmetics
- In 'pre-work' skin protection and barrier creams
- In paint to extend the titanium dioxide (TiO_2) white pigment and modify gloss levels
- For modifying the properties of rubber upon vulcanization
- In adhesives to modify rheology
- In organic farming as a spray applied to crops to deter insect damage, and in the case of apples, to prevent sun scald
- As whitewash in traditional stone masonry homes in Nepal (the most common method is to paint the upper part with white kaolin clay and the middle with red clay; the red clay may extend to the bottom, or the bottom may be painted black)

- As a filler in Edison Diamond Discs
- As a filler to give bulk, or a coating to improve the surface in papermaking
- As an indicator in radiological dating since kaolinite can contain very small traces of uranium and thorium
- To soothe an upset stomach, similar to the way parrots (and later, humans) in South America originally used it (more recently, industrially-produced kaolinite preparations were common for treatment of diarrhea; the most common of these was kaopectate, which abandoned the use of kaolin in favor of attapulgiteand then (in the United States) bismuth subsalicylate (the active ingredient in Pepto-Bismol))
- For facial masks or soap (known as "White Clay")
- As adsorbents in water and wastewater treatment
- To induce blood clotting in diagnostic procedures, e.g. Kaolin clotting time
- In its altered metakaolin form, as a pozzolan; when added to a concrete mix, metakaolin accelerates the hydration of Portland cement and takes part in the pozzolanic reaction with the portlandite formed in the hydration of the main cement minerals (e.g. alite).
- In its altered metakaolin form, as a base component for geopolymer compounds.

Mica

Mica is any of a group of hydrous potassium, aluminum silicate minerals. It is a type of phyllosilicate, exhibiting a two-dimensional sheet or layer structure. Among the principal rock-forming minerals, micas are found in all three major rock varieties—igneous, sedimentary, and metamorphic.

Mica belongs to a very important and rather large group of minerals that are highly suitable for several applications. However, its advanced properties make it highly suitable for use in various places. These are:

- Physical: Mica is translucent, easily split into thin films along its cleavage, optically flat, colorless in thin sheets, elastic and incompressible.
- Chemical: It is a compound hydrous silicate of aluminum, which also contains iron, magnesium, potassium, sodium fluorine, lithium and also few traces of numerous other elements. It is constant and entirely static to the action of water, acids (except for hydrofluoric and concentrated sulphur), alkalies, conventional solvents, bases, and oil. It remains almost unchanged by atmospheric action.
- Electrical: Mica has the exclusive combination of uniform dielectric steadiness, capacitance stability, enormous dielectric power, high Q factor and lower power loss, high electrical resistance and low temperature coefficient. It is highly regarded for its resistances to arc and corona discharge without causing any lasting injury.
- Thermal: It is highly fire proof, incombustible, non-flammable, infusible, and also can resist temperatures of up to 1000 degrees Celsius/1832 degrees Fahrenheit. However this

depends on the type and variety of Mica used. It has excellent thermal stability, lower heat conductivity, and can be easily exposed to high temperatures without visible effect.

- Mechanical: Mica is highly tough, having high tensile strength, elastic, and along with being flexible. It has immense compression power and can be machined, die-punched, or hand cut.

Uses of Mica

1. It is used in paints as a pigment extender and also helps to brighten the tone of coloured pigments.
2. It is a superior insulator. So in the electrical industry it is used as thermal insulation, and in electronic equipment as electrical insulators.
3. Its shiny and glittery appearance makes it ultimate for toothpaste and cosmetics.
4. It is invariably used for fillers, extenders along with providing smoother uniformity, improving workability and prevents cracking.
5. It can be used as an insulator in home attics, concrete blocks and also poured into open top walls.
6. It can also be added to grease to increase its durability and giving it a better surface.
7. It can also be used as a soil conditioner particularly in potting soil mixes and in gardening plots.

Kyanite

Blue kyanite crystals: A very common habit of kyanite is blue bladed crystals.

Kyanite is a mineral found mainly in metamorphic rocks. It most often forms from the high-pressure alteration of clay minerals during the metamorphism of sedimentary rocks. It is found in the schists and gneisses of regionally metamorphosed areas and less often in quartzite or eclogite.

Kyanite's typical habit is a bladed crystal, although it sometimes occurs as radiating masses of crystals. Kyanite is often associated with other metamorphic minerals such as garnet, staurolite, and corundum.

Radiating kyanite: Sometimes kyanite occurs as radiating masses of crystals such as this specimen from Petaca, New Mexico. Specimen is about 4 inches (ten centimeters) across

Kyanite's Unusual Hardness

Kyanite specimens have a variable hardness. The long crystals have a Mohs hardness of about 4.5 to 5 if tested parallel to the length of a crystal, and a hardness of 6.5 to 7 if tested across the short dimension of a crystal. The mineral was once commonly called «disthene" which means "two strengths."

Physical Properties of Kyanite	
Chemical Classification	Silicate
Color	Blue, white, gray, green, colorless
Streak	White, colorless
Luster	Vitreous, pearly
Diaphaneity	Transparent to translucent
Cleavage	Perfect in two directions, faces sometimes striated
Mohs Hardness	Kyanite often occurs in long, bladed crystals. These have a hardness of 4.5 to 5 along the length of the crystals and 6.5 to 7 across the width of the crystals.
Specific Gravity	3.5 to 3.7
Diagnostic Properties	Color, cleavage, bladed crystals
Chemical Composition	Al_2SiO_5
Crystal System	Triclinic
Uses	Ceramics, gemstones

Polymorphs of Al_2SiO_5

Three minerals have a chemical composition of Al_2SiO_5. These are kyanite, andalusite, and sillimanite. Kyanite is the high-pressure polymorph, sillimanite forms at high temperature, and andalusite is the low-pressure polymorph.

Kyanite porcelain sink: Kyanite is used in the porcelain of sanitary fixtures

Many Industrial uses of Kyanite

Kyanite is used to manufacture a wide range of products. An important use is in the manufacture of refractory products such as the bricks, mortars, and kiln furniture used in high-temperature furnaces. For foundries, the molds that are used for casting high-temperature metals are often made with kyanite.

Kyanite is also in products used in the automotive and railroad industries where heat resistance is important. Mullite, a form of calcined kyanite, is used to make brake shoes and clutch facings.

Kyanite spark plug: The porcelain insulator on this spark plug was made with kyanite

Kyanite has properties that make it exceptionally well suited for the manufacture of a high-refractory-strength porcelain - a porcelain that holds its strength at very high temperatures. A familiar use of this type of porcelain is the white porcelain insulator on a spark plug.

Kyanite is also used in some of the more common forms of porcelain, such as those used to make dentures, sinks, and bathroom fixtures.

Expansion of Kyanite when Heated

Kyanite cutting wheel: Kyanite is used as a heat-resistant binding medium in cutting tools and grinding wheels

Kyanite, unlike most other minerals, can expand significantly when heated. Depending upon particle size, temperatures, and heating conditions, kyanite can expand to up to twice its original volume when heated. This expansion is predictable. In the manufacture of certain refractory products, specific amounts of kyanite are added to the raw material (which shrinks during heating) to maintain volume in the finished product.

Dolomite

Dolomite is also known as dolostone and dolomite rock. It is a sedimentary rock which primarily consists of the mineral dolomite. It is found in sedimentary basins worldwide. Dolomite rock is similar to limestone rock. Both dolomite and limestone rocks share the same colour ranges of white-to-gray and white-to-light brown (although other colours such as red, green, and black are also possible). Both the rocks have approximately the same hardness, and they are both soluble in dilute hydrochloric (HCl) acid.

Dolomite is a complex mineral. It is relatively a soft mineral which can be easily crushed to a soft powder. The mineral is an anhydrous carbonate mineral consisting of a double carbonate of calcium (Ca) and magnesium (Mg). It is chemically represented by $CaMg(CO_3)_2$ or $CaCO_3.MgCO_3$. It theoretically contains 54.35 % of $CaCO_3$ and 45.65 % of $MgCO_3$ or 30.41 % of CaO, 21.86 % of MgO, and 47.73 % of CO_2 and has a molecular ratio of 1:1.

However, in nature, dolomite is not available in the above exact proportion. The generally accepted custom today is to say that a true dolomite carries over 40 percent $MgCO_3$. Hence, in commercial language, the rock containing 40% to 45% of $MgCO_3$ is usually called dolomite. Between 40% and 45% of $MgCO_3$, the excess $CaCO_3$ is generally carried in solid solution or as minute calcite inclusions in the dolomite. In limestones with much less Mg, however, there often are still recognizable dolomite crystals.

The dolomitization process results in a slight volume reduction when limestone is converted into dolomite. This can produce a porosity zone in the strata where dolomitization has occurred. These pore spaces can be traps for subsurface fluids like oil and natural gas. This is why dolomite is often a reservoir rock which is sought during the exploration for oil and natural gas. Dolomite can also serve as a host rock for lead, zinc, and copper deposits.

Dolomite deposits are normally associated with limestones. This fact, together with the presence of variable amounts of impurities such as silica (SiO_2), sulphur (S), iron oxides, and alumina (Al_2O_3), has an important bearing on the suitability of dolomite for specific applications.

Dolomite is grouped under flux and construction minerals, It has a number of applications. One of its large applications by volume is in construction where it may substitute for limestone. It is an important raw material for iron and steel, ferro-alloys, glass, fertilizer, and many more industries. Dolomite chips are also used in flooring tiles. For some of the applications, dolomite is specifically valued for its Mg content, and chemical composition is all important. In some other applications, its chemical properties (or degree of whiteness) are important. The MgO content is generally expressed after calcination (removal of CO_2). However, glass industry normally uses pre-calcined MgO content for control purposes and there is a theoretical maximum of around 21.8% of MgO. Dolomite when used in construction is a low value commodity, and hence does not normally lend to long transport distances. However, this is not so in case of high quality industrial dolomite.

Uses of Dolomite

In industry it is being used (i) as a fluxing material, (ii) as a refractory material, and (iii) as a filler material. The major industries where dolomite is being used are iron and steel industry, ferro alloy industry, plastic industry, ceramic industry, glass industry, fertilizer industry, and soap and detergent industry etc.

In iron and steel industry, the main uses of dolomite are (i) as a fluxing material (ii) for protection of refractory lining, and (iii) as a refractory raw material. Dolomite in iron and steel industry is normally used in three forms. These are (i) raw dolomite which is also the natural form of dolomite, (ii) calcined dolomite, and (iii) sintered dolomite. When dolomite is used as a fluxing material then it is used as either raw dolomite or calcined dolomite. When dolomite is used for the protection of refractories, it is used in calcined form and when dolomite is being used as a refractory raw material, it is used in the form of sintered dolomite.

Bentonite

Bentonite is the commercial name of a whole range of natural clays with a high water absorption capacity causing it to expand and swell. Bentonite predominantly consists of montmorillonite: a clay mineral belonging to a class of phyllosilicates called smectites. Bentonites may contain a variety of accessory minerals in addition to montmorillonite. These may include lesser amounts of other clay minerals such as kaolin, mica, illite, as well as non-clay minerals like quartz, feldspar, calcite, and gypsum. Bentonite quality, and, consequently, its applications, depend on whether it contains any of these other minerals.

Its density when dry varies depending on the quality, and may range from 2.2 to 2.8 g/cm^3. Bentonite apparent density, when quarried and piled under natural moisture conditions, ranges from 1.5 to 1.8 g/cm^3. The apparent density of milled bentonite products varies depending on mill fineness, ranging from 0.7 to 0.9 g/cm^3.

Uses

Bentonite used as cat litter

Creating a bentonite slurry for fining after wine pressing

The main uses of bentonite are for drilling mud, binder (e.g. foundry-sand bond, iron ore pelletizer), purifier, absorbent (e.g. pet litter), and as a groundwater barrier. As of around 1990, almost half of the US production of bentonite was used for drilling mud.

Drilling Mud

Bentonite is used in drilling fluids to lubricate and cool the cutting tools, to remove cuttings, and to help prevent blowouts. Much of bentonite's usefulness in the drilling and geotechnical engineering industry comes from its unique rheological properties. Relatively small quantities of bentonite suspended in water form a viscous, shear-thinning material. Most often, bentonite suspensions are also thixotropic, although rare cases of rheopectic behavior have also been reported. At high enough concentrations (about 60 grams of bentonite per litre of suspension), bentonite suspensions begin to take on the characteristics of a gel (a fluid with a minimum yield strength required to make it move). So, it is a common component of drilling mud used to curtail drilling fluid invasion by its propensity for aiding in the formation of mud cake.

Binder

Bentonite has been widely used as a foundry-sand bond in iron and steel foundries. Sodium bentonite is most commonly used for large castings that use dry molds, while calcium bentonite is more commonly used for smaller castings that use "green" or wet molds. Bentonite is also used as a binding agent in the manufacture of iron ore (taconite) pellets as used in the steelmaking industry. Bentonite, in small percentages, is used as an ingredient in commercially designed clay bodies and ceramic glazes.

The ionic surface of bentonite has a useful property in making a sticky coating on sand grains. When a small proportion of finely ground bentonite clay is added to hard sand and wetted, the clay binds the sand particles into a moldable aggregate known as green sand used for making molds in sand casting. Some river deltas naturally deposit just such a blend of clay silt and sand, creating a natural source of excellent molding sand that was critical to ancient metalworking technology. Modern chemical processes to modify the ionic surface of bentonite greatly intensify this stickiness, resulting in remarkably dough-like yet strong casting sand mixes that stand up to molten metal temperatures.

The same effluvial deposition of bentonite clay onto beaches accounts for the variety of plasticity of sand from place to place for building sand castles. Beach sand consisting of only silica and shell grains does not mold well compared to grains coated with bentonite clay. This is why some beaches are much better for building sand castles than others.

The self-stickiness of bentonite allows high-pressure ramming or pressing of the clay in molds to produce hard, refractory shapes, such as model rocket nozzles.

Purification

Bentonites are used for decolorizing various mineral, vegetable, and animal oils. They are also used for clarifying wine, liquor, cider, beer, mead, and vinegar.

Bentonite has the property of adsorbing relatively large amounts of protein molecules from aqueous solutions. Consequently, bentonite is uniquely useful in the process of winemaking, where it is used to remove excessive amounts of protein from white wines. Were it not for this use of bentonite, many or most white wines would precipitate undesirable flocculent clouds or hazes upon exposure to warm temperatures, as these proteins denature. It also has the incidental use of inducing more rapid clarification of both red and white wines.

Absorbent

Bentonite is used in a variety of pet care items such as cat litter to absorb the odor and surround the feces. It is also used to absorb oils and grease.It is also used as a detoxification method.

Groundwater Barrier

The property of swelling on contact with water makes sodium bentonite useful as a sealant, since it provides a self-sealing, low-permeability barrier. It is used to line the base of landfills to prevent migration of leachate, for quarantining metal pollutants of groundwater, and for the sealing of subsurface disposal systems for spent nuclear fuel. Similar uses include making slurry walls, waterproofing of below-grade walls, and forming other impermeable barriers, e.g., to seal off the annulus of a water well, to plug old wells.

Bentonite can also be "sandwiched" between synthetic materials to create geosynthetic clay liners (GCLs) for the aforementioned purposes. This technique allows for more convenient transport and installation, and it greatly reduces the volume of bentonite required. It is also used to form a barrier around newly planted trees to constrain root growth so as to prevent damage to nearby pipes, footpaths and other infrastructure. Farmers use bentonite to seal retention ponds and line canals.

Medical

Bentonite has been prescribed as a bulk laxative, and it is also used as a base for many dermatologic formulas. Granular bentonite is being studied for use in battlefield wound dressings. Bentonite is also sold online and in retail outlets for a variety of indications.

Bentoquatam is a bentonate-based topical medication intended to act as a shield against exposure to urushiol, the oil found in plants such as poison ivy or poison oak.

Bentonite can also be used as a desiccant due to its adsorption properties. Bentonite desiccants have been successfully used to protect pharmaceutical, nutraceutical, and diagnostic products from moisture degradation and extend shelf life. In fact, in the most common package environments, bentonite desiccants offer a higher adsorption capacity than silica gel desiccants. Bentonite complies with the FDA for contact with food and drugs.

In Thai Farming

The application of clay technology by farmers in northeast Thailand, using bentonite clay, has dramatically reversed soil degradation and resulted in greater economic returns, with higher yields and higher output prices. Studies carried out by The International Water Management Institute and partners in 2002–2003 focused on the application of locally sourced bentonite clays to degraded soils in the region. These applications were carried out in structured field trials. Applying bentonite clays effectively improved yields of forage sorghum grown under rain-fed conditions.

Bentonite application also influenced the prices that farmers received for their crops. Production costs are higher, but due to more production and the quality of the food, clay farmers could afford to invest and grow more and better food, compared to nonclay-using farmers.

Bentonite Slurry Walls in Modern Construction

Bentonite slurry walls (also known as diaphragm walls) are used in construction, where the slurry wall is a trench filled with a thick colloidal mixture of bentonite and water. A trench that would collapse due to the hydraulic pressure in the surrounding soil does not collapse as the slurry balances the hydraulic pressure. Forms for concrete, and rebar, can be assembled in a slurry-filled trench, and then have concrete poured into the form. The liquid concrete being denser displaces the less-dense bentonite slurry and causes the latter to overflow from the trench. This displaced bentonite slurry is then channelled to a recycling unit from which it can subsequently be reused in a new trench elsewhere on the construction site.

In addition, because the colloid is relatively impervious to water, a slurry wall can prevent the seepage of groundwater, which is useful in preventing the further spread of groundwater that has been contaminated by toxic material such as industrial waste.

Pottery

The clay from which pottery is made is described as "plastic" if it can be stretched without cracking or "short" or non-plastic if it tends to break up when formed. A small amount of bentonite is often added to make a clay mixture more plastic. This makes the clay more easily formed, by methods such as throwing on a potter's wheel and various hand building techniques. Short clay, such as a porcelain mixture consisting only of kaolinite and feldspar, can be formed only by a limited set of methods such as pressing and molding, that do not put the clay under tension. One problem with adding bentonite is that it typically contains minerals that affect the color of the mix.

Pottery glazes often contain bentonite. The bentonite is added to slow or prevent the settling of the glazes. It can also improve the consistency of application of glazes on porous bisque fired ware. Once a certain amount of glaze water has been absorbed by the bisque ware the bentonite effective clogs the pores and resists the absorption of further water resulting in a more evenly thick coat.

Feldspar

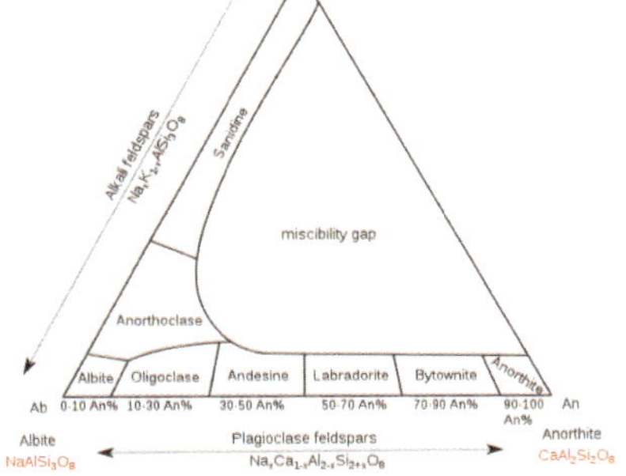

Compositional phase diagram of the different minerals that constitute the feldspar solid solution

Feldspars ($KAlSi_3O_8$ – $NaAlSi_3O_8$ – $CaAl_2Si_2O_8$) are a group of rock-forming tectosilicate minerals that make up about 41% of the Earth's continental crust by weight.

Feldspars crystallize from magma as veins in both intrusive and extrusive igneous rocks and are also present in many types of metamorphic rock. Rock formed almost entirely of calcic plagioclase feldspar is known as anorthosite. Feldspars are also found in many types of sedimentary rocks.

Production and uses

About 20 million tonnes of feldspar were produced in 2010, mostly by three countries: Italy (4.7 Mt), Turkey (4.5 Mt), and China (2 Mt).

Feldspar is a common raw material used in glassmaking, ceramics, and to some extent as a filler and extender in paint, plastics, and rubber. In glassmaking, alumina from feldspar improves product hardness, durability, and resistance to chemical corrosion. In ceramics, the alkalis in feldspar (calcium oxide, potassium oxide, and sodium oxide) act as a flux, lowering the melting temperature of a mixture. Fluxes melt at an early stage in the firing process, forming a glassy matrix that bonds the other components of the system together. In the US, about 66% of feldspar is consumed in glassmaking, including glass containers and glass fiber. Ceramics (including electrical insulators, sanitaryware, pottery, tableware, and tile) and other uses, such as fillers, accounted for the remainder.

In earth sciences and archaeology, feldspars are used for K-Ar dating, argon-argon dating, and luminescence dating.

Fluorite

Fluorite cleavage: Fluorite is the only common mineral with four directions of perfect cleavage. This perfect cleavage combined with the mineral's isometric crystal structure frequently cause it to cleave into perfect octahedrons as shown here. These specimens also show the purple and yellow colors that are typical of fluorite

Fluorite is an important industrial mineral composed of calcium and fluorine (CaF_2). It is used in a wide variety of chemical, metallurgical, and ceramic processes. Specimens with exceptional diaphaneity and color are cut into gems or used to make ornamental objects.

Fluorite is deposited in veins by hydrothermal processes. In these rocks it often occurs as a gangue mineral associated with metallic ores. Fluorite is also found in the fractures and cavities of some limestones and dolomites. It is a very common rock-forming mineral found in many parts of the world. In the mining industry, fluorite is often called «fluorspar.»

Physical Properties of Fluorite	
Chemical Classification	Halide
Color	Typically purple, green, and yellow. Also colorless, blue, red, and black.
Streak	White
Luster	Vitreous
Diaphaneity	Transparent to translucent
Cleavage	Four directions of perfect cleavage
Mohs Hardness	4
Specific Gravity	3.2
Diagnostic Properties	Cleavage, hardness, specific gravity, color
Chemical Composition	CaF_2
Crystal System	Isometric
Uses	Numerous uses in the metallurgical, ceramics, and chemical industries. A source of fluorine, hydrofluoric acid, metallurgical flux. High-clarity pieces are used to make lenses for microscopes, telescopes, and cameras.

Physical Properties of Fluorite

Fluorite is very easy to identify if you consider cleavage, hardness, and specific gravity. It is the only common mineral that has four directions of perfect cleavage, often breaking into pieces with the shape of an octahedron. It is also the mineral used for a hardness of four in the Mohs Hardness Scale. Finally, it has a specific gravity of 3.2, which is detectably higher than most other minerals.

Although color is not a reliable property for mineral identification, the characteristic purple, green, and yellow translucent-to-transparent appearance of fluorite is an immediate visual clue for the mineral.

Fluorite Occurrence

Fluorite crystal mass: An impressive cluster of fluorite crystals from the Berbes Mine, Ribadesella, Asturias, Spain

Most fluorite occurs as vein fillings in rocks that have been subjected to hydrothermal activity. These veins often contain metallic ores which can include sulfides of tin, silver, lead, zinc, copper, and other metals.

Fluorite is also found in the fractures and vugs of some limestones and dolomites. Fluorite can be massive, granular, or euhedral as octahedral or cubic crystals. Fluorite is a common mineral in hydrothermal and carbonate rocks worldwide.

Uses of Fluorite

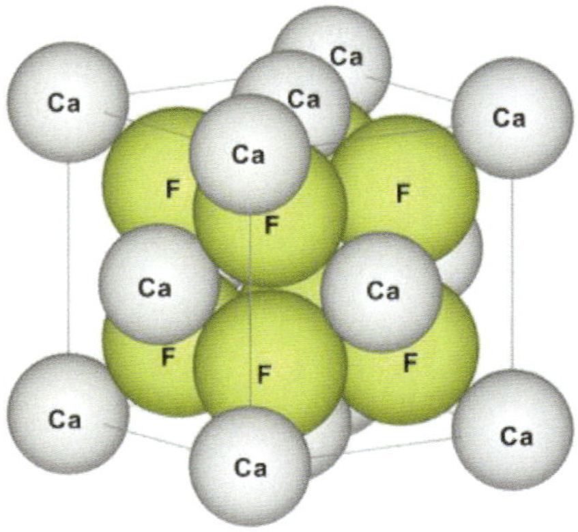

Fluorite unit cell: Illustration showing the relative size and position of fluorine and calcium ions in the isometric unit cell of fluorite

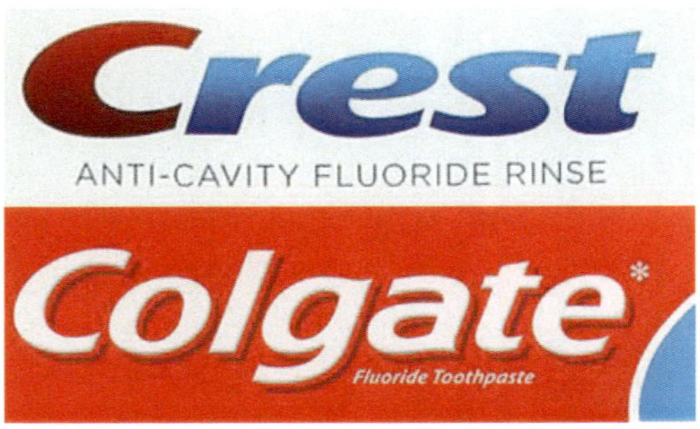

Fluoride products: Most people are familiar with fluoride products used in the prevention of tooth decay. Fluoride is added to drinking water as a systemic fluoride therapy and added to toothpastes, mouthwashes and dental rinse as a topical fluoride therapy. These uses of fluoride have been controversial

Fluorite gemstone: Fluorite can be a beautiful gemstone when faceted. It is mainly a gemstone for collectors because it has a hardness of 4 on the Mohs scale and because it cleaves easily in four directions

Fluorite has a wide variety of uses. The primary uses are in the metallurgical, ceramics, and chemical industries; however, optical, lapidary, and other uses are also important.

Fluorspar, the name used for fluorite when it is sold as a bulk material or in processed form, is sold in three different grades (acid, ceramic, and metallurgical).

Acid Grade Fluorspar

Acid grade fluorspar is a high-purity material used by the chemical industry. It contains over 97% CaF_2. Most of the fluorspar consumed in the United States is acid grade even if it is used in lower grade applications. It is used mainly in the chemical industry to manufacture hydrofluoric acid (HF). The HF is then used to manufacture a variety of products which include: fluorocarbon chemicals, foam blowing agents, refrigerants, and a variety of fluoride chemicals.

Ceramic Grade Fluorspar

Ceramic grade fluorspar contains between 85% and 96% CaF_2. Much of this material is used in the manufacture of specialty glass, ceramics, and enamelware. Fluorspar is used to make glazes and surface treatments that produce hard glossy surfaces, opalescent surfaces, and a number of other appearances that make consumer glass objects more attractive or more durable. The non-stick cooking surface known as Teflon is made using fluorine derived from fluorite.

Metallurgical Grade Fluorspar

Metallurgical grade fluorspar contains between 60 and 85% CaF_2. Much of this material is used in the production of iron, steel, and other metals. Fluorspar can serve as a flux that removes impurities such as sulfur and phosphorous from molten metal and improves the fluidity of slag. Between 20 and 60 pounds of fluorspar is used for every ton of metal produced. In the United States many metal producers use fluorspar that exceeds metallurgical grade.

Optical Grade Fluorite

Specimens of fluorite with exceptional optical clarity have been used as lenses. Fluorite has a very low refractive index and a very low dispersion. These two characteristics enable the lens to produce extremely sharp images. Today, instead of using natural fluorite crystals to manufacture these lenses, high-purity fluorite is melted and combined with other materials to produce synthetic "fluorite" lenses of even higher quality. These lenses are used in optical equipment such as microscopes, telescopes, and cameras.

Lapidary Grade Fluorite

Specimens of fluorite with exceptional color and clarity are often used by lapidaries to cut gemstones and make ornamental objects. High-quality specimens of fluorite make beautiful faceted stones; however, the mineral is so soft and cleaves so easily that these stones are either sold as collector's specimens or used in jewelry that will not be subjected to impact or abrasion. Fluorite is also cut and carved into ornamental objects such as small figurines and vases. These are often treated with a coating or impregnation to enhance their stability and protect them from scratches.

Baryte

Barite or baryte is a mineral containing barium sulfate. Its name is originally derived from the Greek word "barys" meaning "heavy". This mineral belongs to the Barite Mineral Group along with anglesite, celestine and anhydrite. It has quite a beautiful appearance. This mineral ore is also known as Heavy Spar, Barytine, Schwerspath, Barytite and Baryte. It is the main source of barium. Aggregates of barite and sand is called rose rock which appears reddish in due to its iron content. It's commonly known as barite rose.

Barite Characteristic

The unusually high specific gravity of this light colored mineral is the main distinguishing feature. It can also be identified by its insolubility in acid. It can be distinguished from anhydrite and celestite by its orange fluorescence and the green flames that results in flame tests. Its crystal shape, cleavages and lack of fluorite fluorescence differentiate it from fluorites.

Barite Composition

This barium sulfate mineral contains:

- Barium oxide (BaO_2): 65.7%
- Sulfur trioxide (SO_3): 34.3%

Barite

Barite Chemical Formula

The chemical formula for this mineral is $BaSO_4$.

Barite Properties

Here are some basic properties of this material:

- Appearance: It has a transparent crystalline appearance.
- Color: Its color may vary from colorless to white and light shades of yellow, blue, brown and grey.
- Molecular Weight: The molecular weight of this material is 233.4 grams per mole.

- Density: The density of Barite is 4.48 g/ cm^3.
- Solubility: It is almost insoluble in water.
- Crystal System: It has orthorhombic crystal system.
- Crystal Habit: Tabular parallel to base, fibrous, nodular to massive.
- Fracture: It has irregular and uneven fracture.
- Hardness: The hardness of this mineral ranges between 3 and 3.5 on Mohs scale.
- Luster: Barite crystals have vitreous pearly luster.
- Specific Gravity: The specific gravity of Barite is somewhere between 4.3 and 5.
- Streak: It has white streak.
- Tenacity: The tenacity of this crystalline mineral is brittle.
- Diaphaneity: It has transparent to opaque diaphaneity.
- Optical Properties: Biaxial positive.
- Refractive Index: $n\alpha$ = 1.634–1.637; $n\beta$ = 1.636–1.638; $n\gamma$ = 1.646–1.648.
- Birefringence: 0.012.
- Fusibility: It has fusibility of 4. It produces yellowish-green barium flames.

Barite Uses

This mineral ore has many important industrial and commercial uses such as:

- A major amount (77%) of the total Barite found is used as weighting agent for drilling muds. The non-magnetic properties of this material make it more useful for this purpose as it does not interfere with the magnetic measurements of the borehole during the whole process.
- Blue, black, gray or brown Barite is widely used for petroleum well drilling. The applicable Barite color is selected depending on the ore body.
- This is the main ore of barium metal.
- It is also used as paint and plastic filler.
- It is applied for sound reduction in various engine compartments.
- This mineral is used for making smooth and corrosion resistant coat of trucks and automobiles.
- It is used in radiation shielding cements and glass ceramics.

- This mineral also has medical applications. It is used for a barium meal or an upper gastro-intestinal series before doing a contrast CAT scan.
- This mineral was used for producing barium hydroxide for the purpose of sugar refining.
- Historically it was also used as a pigment for paper, textiles, and paint.

Garnet

Gem garnets: Most people think that garnet is a red gemstone. However, garnet occurs in a wide variety of colors. Clockwise from the top left: red almandine (Madagascar), green tsavorite (Tanzania), yellow mali (Mali), orange spessartite (Mozambique), pink malaya (Tanzania), green merelani mint (Tanzania), red pyrope (Ivory Coast), green demantoid (Namibia), purple rhodolite (Mozambique), and orange hessonite (Sri Lanka). Seven out of eight of the garnets above are from Africa, the relatively new source of spectacular garnets

Garnet is the name used for a large group of rock-forming minerals. These minerals share a common crystal structure and a generalized chemical composition of $X_3Y_2(SiO_4)_3$. In that composition, "X" can be Ca, Mg, Fe^{2+} or Mn^{2+}, and "Y" can be Al, Fe^{3+}, Mn^{3+}, V^{3+} or Cr^{3+}.

These minerals are found throughout the world in metamorphic, igneous, and sedimentary rocks. Most garnet found near Earth's surface forms when a sedimentary rock with a high aluminum content, such as shale, is subjected to heat and pressure intense enough to produce schist or gneiss. Garnet is also found in the rocks of contact metamorphism, subsurface magma chambers, lava flows, deep-source volcanic eruptions, and the soils and sediments formed when garnet-bearing rocks are weathered and eroded.

Most people associate the word "garnet" with a red gemstone; however, they are often surprised to learn that garnet occurs in many other colors and has many other uses. In the United States, the major industrial uses of garnet in 2012 were waterjet cutting (35%), abrasive blasting media (30%), water filtration granules (20%), and abrasive powders (10%).

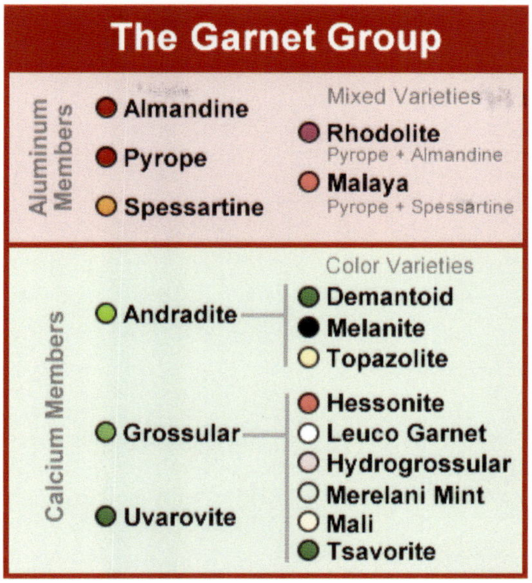

The Garnet Group: This chart summarizes the members of the garnet group that are most important as gemstones. The aluminum garnets are normally red in color with a higher specific gravity and hardness. The calcium members are usually green in color and have a lower hardness.

Physical Properties of Garnet	
Chemical Classification	Silicate
Color	Typically red, but can be orange, green, yellow, purple, black, or brown. Blue garnets are extremely rare.
Streak	Colorless
Luster	Vitreous
Diaphaneity	Transparent to translucent
Cleavage	None
Mohs Hardness	6.5 to 7.5
Specific Gravity	3.5 to 4.3
Diagnostic Properties	Hardness, specific gravity, isometric crystal form, lack of cleavage
Chemical Composition	General formula: $X_3Y_2(SiO_4)_3$
Crystal System	Isometric
Uses	Waterjet cutting granules, abrasive blasting granules, filtration granules, abrasive grits and powders, gemstones

Garnet Physical and Chemical Properties

The most commonly encountered minerals in the garnet group include almandine, pyrope, spessartine, andradite, grossular, and uvarovite. They all have a vitreous luster, a transparent-to-translucent diaphaneity, a brittle tenacity, and a lack of cleavage. They can be found as individual crystals, stream-worn pebbles, granular aggregates, and massive occurrences. Their chemical composition, specific gravity, hardness, and colors are listed below.

Garnet Minerals				
Mineral	Composition	Specific Gravity	Hardness	Colors
Almandine	$Fe_3Al_2(SiO_4)_3$	4.20	7 - 7.5	red, brown
Pyrope	$Mg_3Al_2(SiO_4)_3$	3.56	7 - 7.5	red to purple
Spessartine	$Mn_3Al_2(SiO_4)_3$	4.18	6.5 - 7.5	orange to red to brown
Andradite	$Ca_3Fe_2(SiO_4)_3$	3.90	6.5 - 7	green, yellow, black
Grossular	$Ca_3Al_2(SiO_4)_3$	3.57	6.5 - 7.5	green, yellow, red, pink, clear
Uvarovite	$Ca_3Cr_2(SiO_4)_3$	3.85	6.5 - 7	green
The compositions listed above are for end members of several solid solution series. There are a number of other garnet minerals that are less frequently encountered and not as important in industrial use. They include goldmanite, kimzeyite, morimotoite, schorlomite, hydrogrossular, hibschite, katoite, knorringite, majorite, and calderite.				

There are a variety of different types of garnet, and each has a different chemical composition. There are also solid solution series between most of the garnet minerals. This wide variation in chemistry determines many of their physical properties. As an example, the calcium garnets generally have a lower specific gravity, a lower hardness and are typically green in color. In contrast, the iron and manganese garnets have a higher specific gravity, a greater hardness and are typically red in color.

Garnet as an Industrial Mineral

Garnet abrasive: Garnet granules that have been crushed and size-graded for use as abrasive, cutting, and filter media. They are used in waterjet cutting, “sand” blasting, sandpaper, water filtration, and a number of other uses. Almandine is the hardest garnet and also the most abundant. It is the garnet of choice for most abrasive applications.

Garnet Abrasives

The first industrial use of garnet was as an abrasive. Garnet is a relatively hard mineral with a hardness that ranges between 6.5 and 7.5 on the Mohs Scale. That allows it to be used as an effective abrasive in many types of manufacturing. When crushed, it breaks into angular pieces that provide sharp edges for cutting and sanding. Small granules of uniform size are bonded to paper to produce a reddish color sandpaper that is widely used in woodworking shops. Garnet is also crushed, screened to specific sizes, and sold as abrasive granules and powders. In the United States, New York and Idaho have been important sources of industrial garnet for abrasives.

Waterjet Cutting

Garnet crystal: Almandine, a variety of garnet from River Valley, Ontario, Canada. This specimen is a nice euhedral crystal approximately 2 inches (5 centimeters) across. These types of crystals are often weathered out of a garnet-bearing mica schist and are transported by streams

The largest industrial use of garnet in the United States is in waterjet cutting. A machine known as a waterjet cutter produces a high-pressure jet of water with entrained abrasive granules. When these are directed at a piece of metal, ceramic, or stone, a cutting action can occur that produces very little dust and cuts at a low temperature. Waterjet cutters are used in manufacturing and mining.

Abrasive Blasting

Garnet granules are also used in abrasive blasting (commonly known as "sand blasting"). In these processes, a tool propels a stream of abrasive granules (also known as "media") against a surface using a highly pressurized fluid (usually air or water) as a propellant. Abrasive blasting is done in order to smooth, clean, or remove oxidation products from metals, brick, stone, and other materials. It is usually much faster than sanding by hand or with a sanding machine. It can clean small and intricate surfaces that other cleaning methods would miss. Abrasives of various hardnesses can be used to clean a surface of greater hardness, without damaging the surface.

Filtration

Garnet granules are often used as a filter media. Small garnet particles are used to fill a container through which a liquid flows. The pore spaces of the garnet are small enough to allow passage of the liquid but are too small to allow passage of some contaminant particles, which are filtered from the flow. Garnet is suited for this use because it is relatively inert and has a relatively high specific gravity. Garnet granules, crushed and graded to about 0.3 millimeters in size, can be used to filter out contaminant particles as small as a few microns in diameter. Garnet's high specific gravity and high hardness reduce bed expansion and particle abrasion during backflushing.

Almandine garnet: Almandine, a variety of garnet from Lount Township, Ontario, Canada. This is a granular specimen approximately 11.4 centimeters across

Olivine

Olivine is the name of a group of rock-forming minerals that are typically found in mafic and ultramafic igneous rocks such as basalt, gabbro, dunite, diabase, and peridotite. They are usually green in color and have compositions that typically range between Mg_2SiO_4 and Fe_2SiO_4. Many people are familiar with olivine because it is the mineral of a very popular green gemstone known as peridot.

Physical Properties of Olivine

Olivine is usually green in color but can also be yellow-green, greenish yellow, or brown. It is transparent to translucent with a glassy luster and a hardness between 6.5 and 7.0. It is the only common igneous mineral with these properties. The properties of olivine are summarized in the table:

Physical Properties of Olivine	
Chemical Classification	Silicate
Color	Usually olive green, but can be yellow-green to bright green; iron-rich specimens are brownish green to brown
Streak	Colorless
Luster	Vitreous
Diaphaneity	Transparent to translucent
Cleavage	Poor cleavage, brittle with conchoidal fracture
Mohs Hardness	6.5 to 7
Specific Gravity	3.2 to 4.4
Diagnostic Properties	Green color, vitreous luster, conchoidal fracture, granular texture
Chemical Composition	Typically $(Mg, Fe)_2SiO_4$. Ca, Mn, and Ni rarely occupy the Mg and Fe positions.
Crystal System	Orthorhombic
Uses	Gemstones, a declining use in bricks and refractory sand

Pallasite peridot: This is one of the most incredible gemstones. It is a piece of gem-quality olivine (peridot) from a pallasite meteorite, and it has been faceted into a wonderful little gemstone. This may be the most scarce gem material on Earth - but it actually originated in space. This stone is 2.85 millimeters in diameter and weighs about ten points

Uses of Olivine

Olivine is a mineral that is not often used in industry. Most olivine is used in metallurgical processes as a slag conditioner. High-magnesium olivine (forsterite) is added to blast furnaces to remove impurities from steel and to form a slag.

Olivine has also been used as a refractory material. It is used to make refractory brick and used as a casting sand. Both of these uses are in decline as alternative materials are less expensive and easier to obtain.

References

- Deer, W.A.; Howie, R.A.; Zussman, J. (1992). An introduction to the rock-forming minerals(2 ed.). Harlow: Longman. ISBN 0-582-30094-0.
- Wollastonite-powder-calcium-metasilicate: reade.com, Retrieved 10 April 2018
- Thierry Adatte, Hassan Khozyem, Jorge E. Spangenberg, Bandana Samant & Gerta Keller (2014). “Response of terrestrial environment to the Paleocene-Eocene Thermal Maximum (PETM), new insights from India and NE Spain”. Rendiconti della Società Geologica Italiana. doi:10.3301/ROL.2014.17.
- What-are-the-uses-of-Mica: quora.com, Retrieved 30 June 2018
- Rowe, Aaron (24 April 2008). “Nanoparticles help gauze stop gushing wounds”. Wired.com. Archived from the original on 6 July 2009. Retrieved 2009-08-05.
- Kyanite, minerals: geology.com, Retrieved 25 June 2018
- Noble, A. D.; Gillman, G. P.; Nath, S.; Srivastava, R. J. (2001). “Changes in the surface charge characteristics of degraded soils in the wet tropics through the addition of beneficiated bentonite”. Australian Journal of Soil Research. 39 (5): 991. doi:10.1071/SR00063.
- Dolomite-a-useful-mineral: ispatguru.com, Retrieved 11 July 2018
- Young, Anthony (1980). Tropical soils and soil survey. Cambridge Geographical Studies. 9. CUP Archive. p. 132. ISBN 0-521-29768-0.
- What-is-bentonite: bentonite.it, Retrieved 31 March 2018
- Diamond, Jared M. (1999). “Evolutionary biology: Dirty eating for healthy living”. Nature. Nature. 400 (6740): 120–121. Bibcode:1999Natur.400..120D. doi:10.1038/22014. PMID 10408435.

Chapter 3

Optical Mineralogy

The study of rocks and minerals by an analysis of their optical properties is known as optical mineralogy. Some of the properties measured are refractive index, birefringence, extinction angle, etc. This chapter provides an overview of optical mineralogy and examines the fundamentals of petrographic microscope and thin section.

Optical mineralogy is the optical microscopic study of minerals for identification and also for study of properties. This identification of mineralogical compositions of different rock-types and study of texture, structure etc. in turn is useful in deciphering the information about their origin and evolution.

Optical mineralogy mainly deals with three things:

1. Polarizing microscope (petrographical microscope).
2. Visual light (white light).
3. Thin section.

Polarizing Microscope

Polarized microscope is of possible interest to many sciences that are concerned with crystalline materials such as: geology, mineralogy, crystallography, materials science, biology, forensic science.

Polarizing microscope has four differences as compared to ordinary microscope, they are:

1. It has rotating circular stage which graduated in to 360 degrees.
2. It has Polarizer and analyzer plates: In modern microscopes polaroid sheets are used as polarized and analyzer but in old microscopes Nicol prism or Ahern prism are used the polarizer and analyzers used in our microscopes are made of Polaroid sheets which consist of a sheet of cellulose fed with crystals of quinine iodosulfate which is absorbing light in one direction but transmit in the other direction.
3. It has Accessory plates: such as mica plate, gypsum plate and quartz wedge.
4. It has Bertrand lens: located between analyzer and ocular, which brings the image of interference fiure in to the focal plane of the ocular.
5. Objectives : they are used for magnification of the original object on the stage, two numbers are engraved on the tube of objective lenses:
   - Magnification of the original object e.g: (1x, 3x, 10 x, 40 x and 100 x) the objective lens in the diagram has magnification power of (10) time more than original object on the stage.

- Numerical apertures (N.A)

  N.A = sin u

  U = angular operator/2

  100000 N.A = lines separated/1 square inch

  So if N.A = 0.25 it means that 25000 lines can be separated in one square inch.

Visual Light

The vibration light represent limited band of wave length with in electromagnetic spectrum, ranging from 3900 to 7700 A (angstrom). If light of all wave length simultaneously reach the human eyes, the light is interpreted by the brain as white light.

But light with one wave length called mono chromatic light, for example, sodium vapor lamp is a source of mono chromatic light with wave length of 5890 A but the tungsten lamp is a source of white light (ordinary light).Traverse wave terminology and shape in which polarization and interference occur. Only sound wave is longitudinal which means that the vibration direction is parallel to direction of propagation in which polarization and interference do not occur.

Thin section

It is a fragment of rock or mineral mechanically reduced to thickness of (0.03mm) by grinding and polishing of thickness makes most rocks and minerals transparent which finally ready for optical study.

Procedure for Making Thin Section

A thin section is a 30 μm (= 0.03 mm) thick slice of rock attached to a glass slide with epoxy. Typical thin section slides are 26 mm x 46 mm, although larger ones can be produced. They are generally covered by another glass slide, a cover slip also attached to the rock with epoxy. The epoxy ideally has an index of refraction of 1.54, although our epoxy is slightly higher, perhaps 1.56.

The sections may be left uncovered for chemical analysis on the SEM or electron microprobe. If so, temporary cover slips may be weakly attached with glycerin.

1. Sawing a hand specimen to get a flat chip which has 4 square cm in area.

2. Polishing one of the two surfaces and then mounting on glass slide using Canada balsam.
3. Grinding of the chip, which is mounted on the glass slide, to a thickness close to 0.03mm with (1000) grade carborundum.
4. Then the slide covered with thin cover slide by liquid Canada balsam.
5. During polishing the chip must be examined several times under polarizer microscope to determine the standard thickness of the thin section (0.03mm). This done by the interference color of the quartz or feldspar which they give first order grey when the thickness reaches (0.03mm).

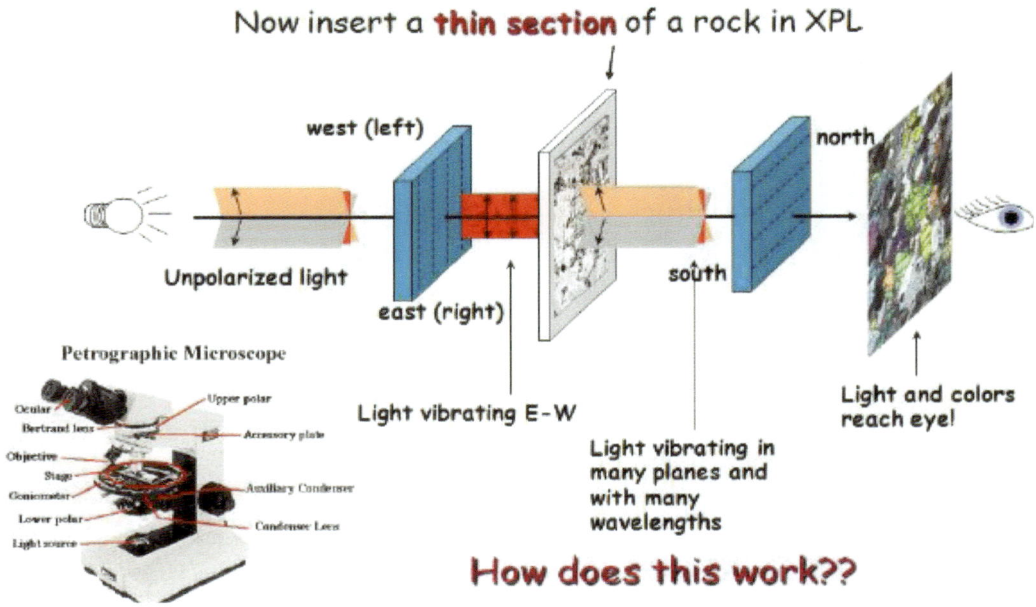

Position (location) of the thin section in the light ray pass from its radiation from the bulb to the eyes of human.

Properties and Nature of Light

The study of the following properties of light are necessary for:

1. Light represents a relatively limited band of wave lengths within the electromagnetic spectrum, this band ranging from 3900 to 7700 armstrom.
2. Light is traversal wave which can undergo: a) interference b) polarization
3. Reflection
4. refraction.

Optical properties relevant to minerals and gems:

- Refractive Index (RI)
- Double Refraction (DR)

- Dispersion and Fire
- Absorption Spectrum.

Refractive Index

The speed of light varies in substances. The speed of light is different in air, water, and other dimensions, including minerals and gemstones. When light travels from one dimension to another dimension, the light bends, or *refracts*, upon entering the second dimension. This phenomenon can be witnessed with a stick protruding from a pond, where the stick appears to "bend" at the water level. This is caused by the difference in the speed of light in air and the speed of light in water. How much the light will bend, or the *angle of refraction*, depends on difference in the speed of light of the two substances. All transparent gemstones refract light, since the speed of light is different in the air than in gemstones.

Every gemstone refracts at a distinct, individual angle. The angle of refraction is directly related the speed of light in the gemstone. The *refractive index* of a gemstone measures the difference between the speed of light in air and the speed of light in the gemstone. This is determined by the gemstone's angle of refraction. Every gemstone has a unique refractive index, meaning every gemstone refracts light at a unique angle.

The refractive index value measures how much slower light travels in the gemstone than in the air. For example, the refractive index of Diamond is 2.42. This means that the speed of light in Diamond is 2.42 times slower than the speed of light in air.

Refractive indices of minerals range from 1.2 to about 3. However, gemstones with a refractive index greater than Diamond (2.42) are either synthetic or are too soft for practical gemstone use. The greater the refractive index of a gemstone, the more brilliant or lustrous it is.

The refractive index of a gemstone is measured with a refractometer, a tool that measures angle of refraction. This tool is used by almost all gemologists and professional jewelers, for it provides simple, inexpensive, and accurate gem identification. However, a refractometer cannot read values greater than 1.86. Gemstones with refractive indices greater than 1.86 can only be tested by placing them in a liquid with a known refractive index, and then calculating the difference in refraction between the liquid and the gem.

Double Refraction

Another optical property, known as double refraction or birefringence, is present in all non-amorphous minerals that do not crystallize in the isometric crystal system. When light rays enter birefringent minerals (minerals with double refraction), the light divides into two rays. The two rays differ in their angle of refraction. Therefore, all birefringent minerals have two refractive indices, one for each ray. The double refraction in most minerals is so weak that it cannot be observed with the naked eye. However, a small number of minerals have a strong double refraction, which is easily seen when the crystal placed over an image appears to "double" the image.

Double refraction is an important guide to gem identification. When viewed through a refractometer, birefringent minerals show two readings – one for each refracted ray of light. Double refraction is a characteristic trait, meaning every specimen of the same gem always has the same double refraction.

Double refraction is measured by the difference of refraction in each light ray. For example, if a gem is placed in a refractometer and shows a double reading of 1.62 and 1.63, its double refraction is .01.

Highly birefringent gemstones, such as Zircon, must be cut in a way where the double refraction is least noticeable in the finished gem, for it otherwise appears blurry.

Dispersion and Fire

Dispersion is the splitting of white light into the colors of the spectrum. This effect is observable in faceted, transparent, colorless gems, where the white light *disperses* in the gem and reflects on its inner surfaces, giving the gem a colorful sparkle. This effect is known as *fire* in the gem trade. Generally, gems with higher refractive indices display greater fire. Diamond has the greatest fire of all true, non-synthetic gems. Room lighting conditions play an important role in fire, for the stronger the light, the more intense the fire appears to be.

The design of the *brilliant cut* was extensively researched to offer a gemstone its maximum amount of fire. For this reason, many transparent gems with high dispersion are faceted with the brilliant cut.

Absorption Spectrum

The *absorption spectrum* describes the spectral wavelengths absorbed by a gem. The chemical structure of each gem allows only certain wavelengths to be absorbed (the rest are reflected or pass through). The wavelengths absorbed into a gem can be detected with an instrument known as a *spectroscope*. Color alone cannot be used to identify a gem, for many gems have identical colors. A spectroscope examines the "true" color of a gem. Examining the absorption spectrum of a gem is one of the most useful and practical methods of gem identification.

Optical Relief

The relief of the mineral is described as the degree to which the mineral grains stand out from the mounting medium. The relief is usually high if the mineral stands out strongly and the indices of the mineral and oil differ roughly at about 0.12 or more. The relief is considered low if the mineral grains do not appear to stand out from the mounting medium and the indices of refraction of the mineral and oil differ from each other roughly about 0.04. The intermediate and the indicator that the indices of oil and mineral differ roughly at about 0.04 to 0.12 is what we called the moderate relief. The ranges should however be used only as approximate guides because the perception of relief is subjective to some degree, and varies depending on the adjustment of the petrographic polarizing microscope.

The mineral is said to have a positive relief when its index of refraction is higher than the oil. It has negative relief if its index is lower than the index of oil. But then, because positive and negative relief looks alike, a method is very much required in order to determine whether the index of refraction of the oil is higher or lower than that of the mineral. There are two convenient methods suggested to use. These are the Becke line method and the oblique illumination method.

Becke Line Test

A band or rim of light called the Becke line is usually visible along the grain boundaries in plane light when these mineral grains in grain mount are slightly out of focus. The location of the rim of light depends on exactly how the microscope is focused. Sometimes the rim of light is found inside and sometimes outside of the grain boundaries. If the aperture is stopped down somewhat and the intermediate power objective, for instance 10X is used, the Becke line is usually most distinct. The Becke line also sometimes appears to move into the material with the higher index of refraction if the focus is raised so that the distance between the sample and the intermediate power objective lens is increased. The production of the Becke line usually involves the lens effect and the internal reflection effect.

The lens effect depends on the observation that most mineral fragments are thinner on the edges than in the middle. That is why they act as crude lenses. The grains act as converging lenses and the light is concentrated toward the center of the grain if the mineral has higher index of refraction than the oil. The grains act as diverging lenses and the light is concentrated in the oil if the grains have a lower index of refraction than the oil.

The internal reflection effect on the other hand depends on the requirement that the edges of the grains must be vertical at some point. The moderately converging light from the condensor impringing on the vertical grain boundary is found either refracted or internally reflected. This is usually depending on the angles of incidence and the indices of refraction. The result of the refraction and internal reflection is to concentrate light into a thin band in the material with the higher index of refractions.

As can be noticed, both the lens effect and the internal reflection effect tend to concentrate light into the material with the higher index of refraction in the area above the mineral grain. Becke line is conveniently considered as the cone of light propagating up from the edges of the mineral. If the index of the mineral is greater than the index of oil, the cone converges above the mineral. However, if the index of the mineral is less than the index of oil, the cone diverges. The Becke line is coincident with the edge of the grain or it may disappear if the microscope is crisply focused on the grain. As the stage is lowered, the Becke line moves toward the material with the higher index of refraction. If the stage is raised, the reversed is also true.

It should be noticed that the Becke line is often paired with a dark band or shadow that moves in the opposite directions as the stage is raised or lowered. The light that might otherwise be present there in the dark bands is reflected or refracted to from the Becke line. That is why the dark bands exist in Becke line.

Oblique Illumination Method

Anyone who has ever examined a coin in direct, bright light will have observed that the relief on the surface of the coin is very difficult to see under these illumination conditions, due to specular reflections from the surface of the coin. On the other hand, when incident light is arranged to "strike" the coin at a low glancing angle, the resulting shadow effect on one side and brightness on the other side (nearer the light) cause the relief detail of the coin to stand out in three-dimensional clarity.

Configuration for Oblique Illumination

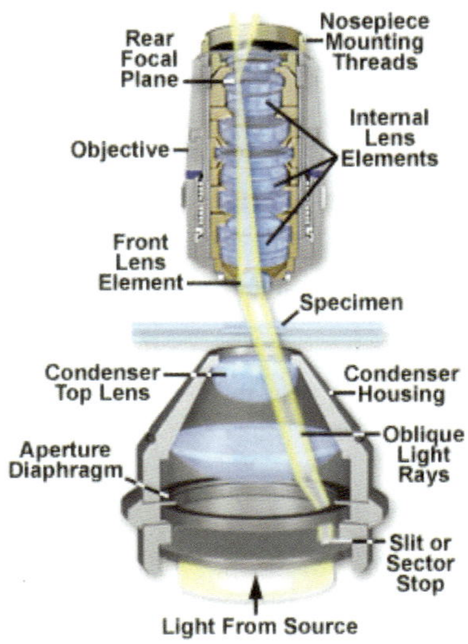

A somewhat similar appearance is produced when examining microscopic specimens with a classical technique known as oblique illumination (sometimes referred to as anaxial illumination). Direct light restricted to a single azimuth of the condenser light cone is allowed to illuminate the specimen from one side only, as illustrated in figure above. The net effect is to reveal details in pseudo-relief from an otherwise almost invisible, colorless specimen. A typical configuration for oblique illumination in compound transmitted light microscopy is presented in figure above. Beneath the condenser lens system is positioned an opaque light stop having a small slit (or sector stop), which is removed from the central optical axis of the microscope. Illuminating light rays from the source pass through the slit and subsequently through the condenser lens element system from a single azimuth that is off-axis. The oblique light rays then illuminate the specimen from a highly specific angle, on one side, and enter the microscope objective through the front lens element near the edge of the aperture (very close to the lens mount). The focused light rays converge at the objective rear focal plane and are directed to the microscope intermediate image plane, where they are observed through the eyepieces.

Oblique lighting has caused the zeroth order to be moved to a position just within the periphery of the objective aperture (this effect can be observed with an eyepiece telescope or Bertrand lens focused at the rear focal plane of the objective). The consequence of oblique illumination is to shift the zeroth order of light passing through the specimen from the center to one peripheral side of the objective front lens element. This shift of the zeroth order to a single side enables one or more additional higher orders of the diffracted light to be included at the rear focal plane of the objective and contribute to formation of the image. Only diffracted orders on a single side of the zeroth order (often termed sidebands) are admitted to the objective and, because of the obliquity of illumination, diffracted orders on the other side miss the objective altogether. This concept is illustrated in figure below utilizing ray traces and a simple lens to represent the microscope objective.

In general, when transparent specimens are illuminated in transmitted brightfield microscopy with a parallel bundle of light rays, the resulting images have poor visibility and lack any significant degree of contrast. This occurs because the sum of the light diffracted by each detail of the specimen is a quarter-wavelength out of phase with direct light passing through the specimen when both are recombined at the image plane. However, the diffracted light emanating from each specimen point contains sidebands on either side of the zeroth order undeviated light, which are offset from the quarter wave displacement by a defined margin. If one sideband of the diffracted light is prevented from reaching the image plane, constructive and destructive interference will occur between the remaining sideband and the direct light to produce a visible image at the intermediate image plane of the microscope. Oblique illumination is an excellent mechanism to achieve the sideband suppression necessary for the image formation process.

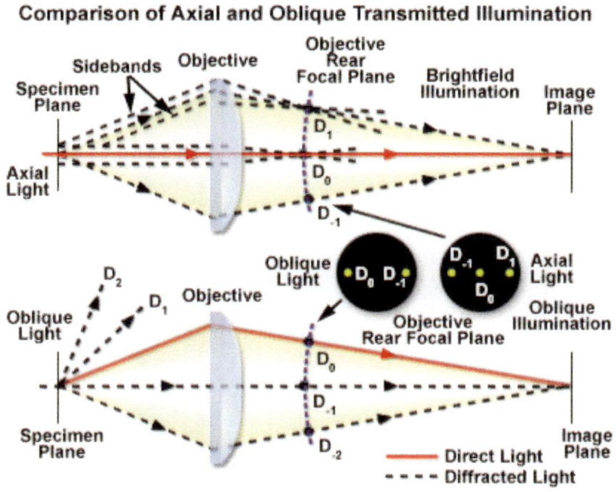

In figure above, the uppermost ray trace system identifies the pathway for axial light rays passing through the specimen and into the microscope optical system for a standard brightfield configuration. Symmetrical components of the diffracted light are admitted into the objective and produce complimentary patterns having no significant interference (either constructive or destructive) with the direct light components at the image plane, leading to images having poor contrast and visibility. With oblique illumination (the lower ray trace system in Figure), several of the diffracted component sidebands are not captured by the objective, but at least one higher order component is admitted along with the zeroth order (the second order sideband (D(-2)) is not included at the objective rear focal plane in the figure to increase the clarity of the illustration). Interference at the image plane produces an image of the specimen having significantly more contrast and visibility. The appearance of diffraction maxima at the objective rear focal plane for both axial and oblique illumination is also presented in the figure. Because only two maxima are necessary for image formation, oblique illumination will produce a visible image.

Oblique illumination is capable of resolving very fine specimen detail that is difficult to distinguish using conventional brightfield techniques. When oblique light rays strike the objective front lens at angle "ob" to the microscope optical axis, and the refractive index of the medium between the condenser front lens and the objective is n, the relationship between resolution, illuminating wavelength (λ), and refractive index can be described as:

$$\text{Resolution (D)} = \lambda/(n \cdot \sin(ob)) + \text{Numerical Aperture (NA)}$$

Some authors equate the term n • sin (ob) with the numerical aperture of illumination, but this is misleading because the equation is not intended to signify a full cone of light emanating from all azimuths, but is restricted to a beam of light produced by anaxial illumination from a single azimuth. In cases where specimen detail is so fine that the zeroth order undiffracted and first order diffracted sideband light are separated by a distance equal to the diameter of the objective aperture, the resolving power is twice as high as observed for axial transmitted illumination and is expressed by the equation:

$$\text{Resolution (D)} = \lambda/2\text{NA}$$

The optical conditions required for the microscope to conform with the equation above (when utilizing oblique illumination) are such that the fine details of specimen periodicity are limited by the resolving power of the objective. In many cases, specimen details that are resolved utilizing brightfield illumination are so severely lacking in contrast that they can scarcely be visualized or imaged. The net result of observing specimens with oblique illumination is often an increase in resolution (over that obtained in brightfield illumination with a closed condenser aperture diaphragm) and also the production of a shadowed, relief-like pseudo three-dimensional appearance in the image of the specimen.

Oblique illumination techniques are ideal for imaging a wide variety of unstained objects such as living cells, crystals, diatoms, and similar transparent or semi-transparent specimens. However, the resulting images must be viewed and interpreted with caution because the diffracted orders from one side have not contributed to the image formation. Potential false structures appearing in the image can seriously limit the usefulness of oblique light to examine and quantitatively describe previously unobserved specimen detail. This fact should always be taken into consideration when observing specimens with oblique illumination.

Condenser with Decenterable Aperture Diaphragm

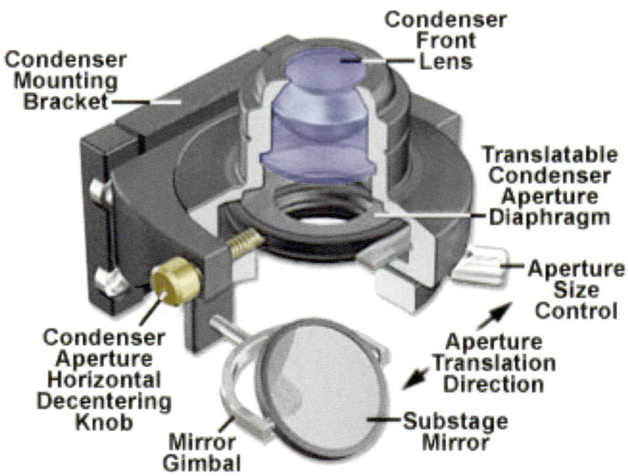

Achieving conditions necessary for oblique illumination, which has been employed to enhance specimen visibility since the dawn of microscopy, can be accomplished by a variety of techniques with a simple transmitted optical microscope. Perhaps the easiest methods are to offset a partially closed condenser iris diaphragm or the image of the light source. In former years, some microscopes were equipped with a condenser having a decenterable aperture iris diaphragm (note that the lens system of the condenser should not be displaced from the center of the microscope optical axis). The device was engineered to allow the entire iris to move off-center in a horizontal plane so that closing the circular diaphragm opening would result in moving the zeroth order to the periphery of the objective rear focal plane. In advanced models, the entire diaphragm was rotatable around the axis of the microscope so that oblique light could be directed toward the specimen from any azimuth to achieve the best desired effect for a given specimen.

In practice, utilizing a partially closed condenser iris diaphragm technique to achieve oblique lighting is plagued with problems similar to those hampering brightfield observation. The results are a general loss of resolution and superimposition of diffraction rings, which surround and confuse the interpretation of minute specimen detail. In addition, Becke lines and other undesirable optical effects originating from regions of the specimen that are not in exact focus complicate the image. To circumvent many of these problems, a combination of oblique illumination at large condenser numerical aperture, coupled to video contrast enhancement, has been utilized as an effective method to generate optical thin sections that strongly resemble those observed with differential interference contrast (DIC) microscopy.

Laterally Displaced Lamp Filament

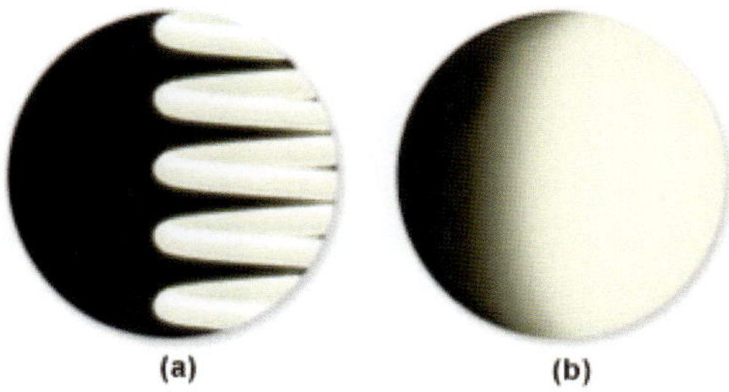

Although oblique illumination may be configured by decentering the condenser aperture, a majority of the condensers currently sold by the microscope manufacturers do not have this capability. An alternative technique involves lateral displacement of the tungsten lamp filament from the microscope optical axis, as illustrated in figure above. The diagram in figure(a) above presents the image of a focused filament, which has been translated to the right by adjustment of the lamp position in the lamphouse at the rear of the microscope. When a diffusion filter is inserted into the light path, the light is spread to fill the condenser aperture with a gradient of illumination having the proper orientation for viewing specimens with the oblique method. Optimum gray tones can be achieved, in most cases, by placing the filament image in a position to cover approximately half of the condenser aperture diaphragm. However, careful adjustment of the filament position and condenser aperture, coupled to the utilization of diffusion screens having differing opacities, should allow the microscopist a significant degree of latitude in order to fine-tune modulation of image contrast.

Sector Stop Configurations

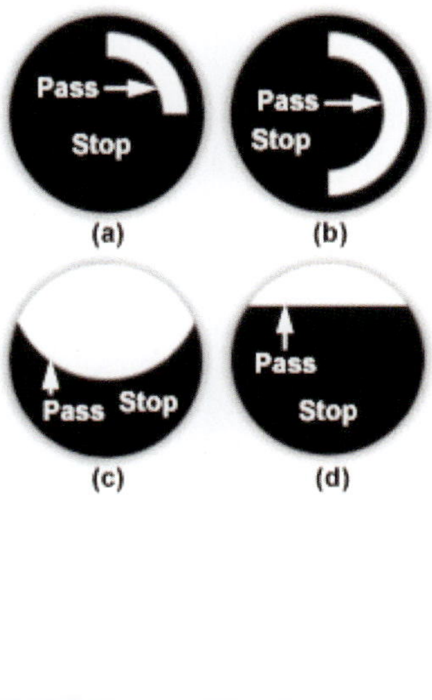

A more commonly employed alternative for establishing variable degrees of oblique illumination is the use of sector stops placed just below the lower lens element and aperture diaphragm of the condenser. Many microscopes are equipped with a filter holder positioned near the bottom of the condenser, providing an ideal location for insertion of a sector stop. Illustrations of several sector stop designs are presented in Figure. In the past, sector stops were available from the microscope manufacturers, but they are increasingly difficult to find today. However, simple stops can be made from a section of opaque paper or cardboard and cut to fit the filter tray or taped to the underside of the condenser.

Each sector stop has a section that blocks light and another area that allows only oblique light to pass through and illuminate the specimen. The size of the stop and pass sections often vary with objective magnification and numerical aperture, and proper size determination is largely an experimental effort on the part of the microscopist to achieve the optimum balance. The sector stop can be rotated within the filter holder to illuminate the specimen from a variety of angles, or (more conveniently) a circular rotating stage can be employed to rotate the specimen through 360 degrees to achieve the desired oblique lighting effect. These suggestions can be employed as a guideline, but the individual microscopist should make final adjustments to both the condenser and sector stop configuration while observing the effects on the specimen image to optimize oblique illumination and enhance contrast and detail.

Oblique illumination can dramatically increase specimen contrast when compared to conventional brightfield techniques, as illustrated by the digital images presented in figure below. The specimen in Figure(a) below is a bovine arterial cell that has been imaged using brightfield illumination with the condenser aperture adjusted for maximum contrast. It is obvious that the specimen is almost invisible and details are very difficult to distinguish in brightfield illumination. In contrast, when the specimen is illuminated obliquely using sector stops placed near the condenser aperture, contrast is dramatically increased and many cellular details, including location of the nucleus and pseudopods, become visible. This particular specimen is difficult to image, regardless of the contrast enhancement technology employed by the microscopist (including phase contrast, differential interference contrast, darkfield, or Hoffman modulation contrast).

A somewhat easier specimen to image is the mouse kidney thick section presented in figures(c) and (d) below. In brightfield illumination, some specimen detail is visible, but specific features are difficult to distinguish and the overall image suffers from a serious lack of contrast. Utilizing sector stops to illuminate the specimen obliquely produces a significant improvement in contrast, as shown in figure(d) below. In this digital image, minute specimen details become visible that were previously difficult to observe in brightfield illumination.

When compared to the darkfield technique, where the specimen is illuminated from all azimuths with highly oblique light, asymmetrical oblique illumination produces images whose character is highly dependent upon the incident angle of illumination. Images produced by oblique illumination are asymmetrical in the sense that edges lying perpendicular to the direction of incident illumination are made visible, while those that lie parallel (or close) to this direction are not. This concept is illustrated in figure below for two specimens at different orientations with respect to the angle of incident oblique illumination. The specimen in figures (a) and (b) below consists of identical viewfields (but rotated by 90 degrees) obtained from a thin single-crystal wafer of lanthanum aluminate, a perovskite that is commonly employed as a substrate for epitaxial thin film deposition of high-temperature superconducting ceramics. Twinning in these crystals hampers confluent thin film formation and can have detrimental effects on the properties of resulting films. The digital image presented in figure(a) below depicts the pseudo-relief generated by twinning domains when the crystal is oriented with the longitudinal twin axis parallel to the oblique incident light rays. In contrast, when the crystal (and twin) axis is rotated by 90 degrees, so that it is perpendicular to the incident rays, the twinning domains become readily apparent. This represents a spectacular display of specimen orientation restrictions on the textural effects observed under oblique illumination.

Contrast Enhancement with Oblique Illumination

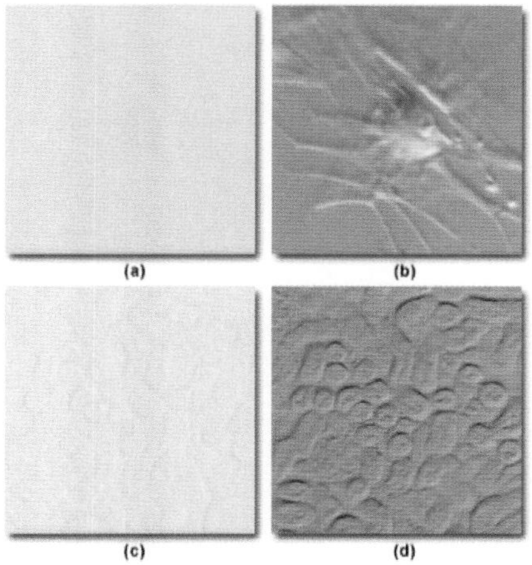

A similar, but less dramatic, result is obtained when observing semi-transparent goat hair specimens with oblique illumination at several orientation angles. When the long axis of the rod-like hair strands is oriented parallel to the incident oblique illumination, structural details in the central portion and edges of the hair fibers is revealed. This detail is absent when the hair fibers are oriented perpendicular to the illumination axis, and a significant difference is observed in the apparent thickness of the fibers between the two orientation angles. Fibers oriented parallel to the incident illumination appear to be much thicker than those oriented perpendicular to the light source. Thus, it is evident that the oblique illumination technique cannot be reliably employed to generate faithful measurement data from images gathered by this method.

Orientation Angle Effects on Specimen Appearance

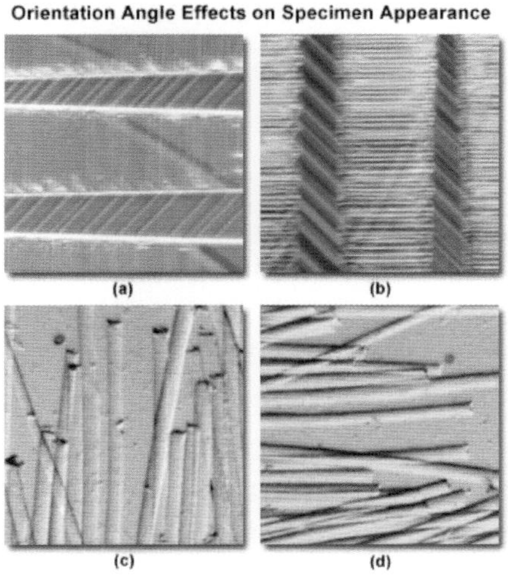

The apparent three-dimensional effect afforded by oblique illumination techniques does not represent the actual specimen geometry or topography, and should not be employed to conduct measurements of specimen dimensions. The true value of the oblique illumination image is in revealing transitions in refractive index or other optical path differences within the specimen that enable the morphology and internal structural arrangement to be more clearly understood. The technique can be applied to a variety of materials that appear nearly invisible or transparent in brightfield illumination and cannot be stained or otherwise chemically or thermally treated to enhance contrast. Study of living organisms and processes such as *in vitro* fertilization, glass or acrylic fibers, chemical crystals, and other unstained materials can be facilitated by the utilization of an easily controlled oblique illumination system.

Refractive Index Determination by Oblique Illumination

Oblique illumination is sometimes utilized as an alternative to the Becke line test to determine whether the refractive index of a specimen is higher or lower than that of the surrounding medium. If the specimen is mounted in a medium of lower refractive index, shading that results from the anaxial illumination will appear on the side opposite to that from which the light enters the specimen, and vice versa, as illustrated in figure below. For both diagrams presented in figure below, two equal sized oblique light rays are depicted entering the specimen through the surrounding medium at the same angle of incidence. At point A on the left-hand diagram, the light is spread over a larger area of the specimen than at point B, so that the area near point A on the specimen appears darker than the area near point B. Under these conditions, one side of the specimen will appear shaded or somewhat darker than the other side when viewed through the microscope eyepieces (A' and B' in the upper left portion of Figure). This is the case when the specimen refractive index is higher than that of the surrounding medium.

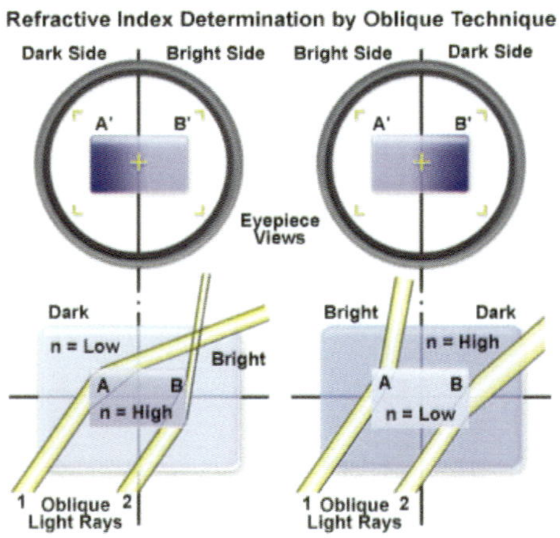

The opposite effect occurs when the specimen has a lower refractive index than that of the surrounding medium. In this case, the shaded or darker side of the specimen will be on the side that is nearest to the oblique light sector stop. When the specimen and the surrounding medium have identical refractive indices, then the specimen will be transparent (or invisible) and will have no refractive effects on the oblique illumination. The sensitivity of this refractive index determination technique is highly dependent on the condenser focal length, the iris diaphragm position, and the geometry of sector stops (if employed). In general, the best results are obtained when the condenser is carefully focused and an even field of illumination is achieved.

Oblique Illumination in Reflected Light

Reflected or incident light oblique illumination, although largely displaced by differential interference contrast and darkfield techniques, is often useful for examination of metals and other specimens in the field of metallurgy. A comparison between the basic configurations for transmitted and reflected light oblique illumination is presented in figure below. The diagram in figure(a) below illustrates the principal optical axes for axial (brightfield), oblique, and darkfield transmitted illumination. A simple lens simulates the microscope objective and yellow arrows represent hypothetical light rays oriented at the respective angles necessary to achieve the desired illumination mode. The angle θ refers to one-half the objective angular aperture. Light rays that are directed along the microscope optical axis are representative of axial illumination, while those lying between the axis and the limiting angle θ are captured by the objective front lens and form oblique illumination. Light rays having an incident oblique angle larger than θ do not enter the objective and represent darkfield illumination.

Oblique Illimination in Transmitted and Reflected Light

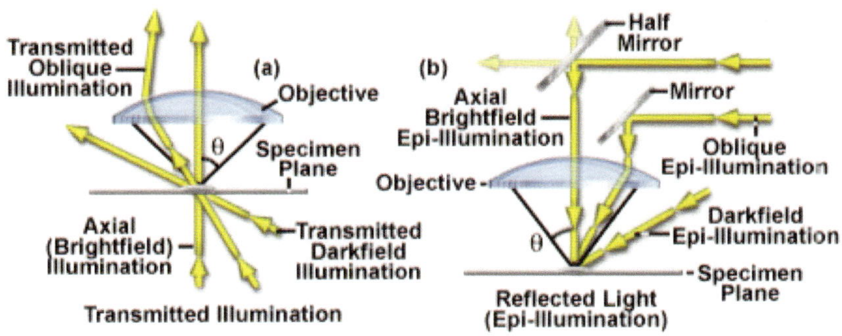

Principal light paths for reflected light or epi-illumination are presented in Figure(b) above. Axial (brightfield) epi-illumination proceeds in a direction parallel to the microscope optical axis from the vertical illuminator to the half mirror, where it is reflected through the objective lens system and onto the specimen. In contrast, oblique epi-illumination is reflected onto the specimen by a 45-degree prism or mirror mounted at the edge of the objective. These light rays pass through the periphery of the objective front lens element and strike the specimen obliquely, but are reflected, refracted, and diffracted back through the objective. As is the case in figure(a) below, the value θ represents the one-half objective angular aperture through which all rays of axial and oblique illumination are directed. Light rays having a higher angle represent darkfield epi-illumination, which can originate from an external light source or an annular mirror or lens system that surrounds the objective. In figure below, only one imaging ray is illustrated for each illumination mode, but in reality, these light rays will propagate in all directions.

Oblique reflected or epi-illumination is suitable for the examination of periodic or linear structures on the surface of metals, ceramics, glasses, polymers, and other opaque specimens. For optimum results, the azimuth of illumination should be positioned at a 90-degree angle to the longitudinal direction of the structures under observation. The technique is very useful for examination of relief in surface structures (especially at lower magnifications), and can be utilized to demonstrate whether a structure on the surface of the specimen is a cavity or protrusion by determining the position of shadows relative to the illumination azimuth. In reflected oblique illumination, specimens should be examined through the complete range of inclination angles and azimuth directions available with the instrumentation to obtain the best possible analysis. If possible, a rotating circular stage should be employed, and the specimen rotated through a 360-degree angle to capture all essential surface detail. It should be noted that because oblique illumination can produce artifacts such as color fringes and directional contrast, the microscopist should compare oblique images to those obtained in brightfield illumination before attempting to interpret the results.

Advanced Oblique Illumination Techniques

Several microscopy techniques related to oblique illumination, including phase contrast and Hoffman modulation contrast, take advantage of fundamental optical parameters of the microscope. Specifically, these methods are based on the fact that the condenser and objective lens apertures are in principal conjugate planes and that the illuminating beam arising from each point of the condenser aperture is diffracted at a different angle (according to its spatial frequency) by the specimen. Another basic concept, utilized by these techniques, is the fact that the objective rear aperture is the Fourier optical plane of the microscope.

In Hoffman modulation contrast microscopy, a sophisticated derivative of the oblique illumination technique, the condenser aperture is fitted with a slit mask having the slit positioned near the edge of the aperture. The objective rear focal plane contains a second, complementary mask, termed a modulator, which consists of three zones having differing degrees of transmittance. Oblique illumination passing through the condenser slit is projected onto the central zone of the modulator and forms an image of the slit in this zone. The central zone has an optical density allowing only 15 percent of the light entering the zone to be transmitted through the modulator, while a darker (but smaller) zone transmits only 1 percent of the light. The largest area of the modulator plate comprises the third zone, which is transparent, and passes essentially all of the incident light (minus that lost due to reflection and absorption). The objective modulator plate (mask) thus attenuates the zeroth-order light undiffracted by the specimen (that passing through the zone of 15 percent transmission) while the light diffracted toward the dark sector (1 percent transmission) is blocked. The light diffracted by the specimen away from the dark sector of the mask passes unchanged, resulting in a shadow-cast image similar in appearance to that produced by differential interference contrast, which enhances gradients of refractive indices or of optical path differences in the specimen.

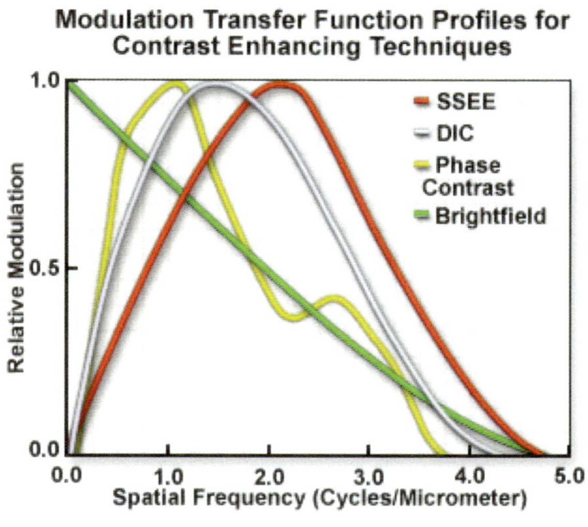

A related technique, known as single-sideband edge enhancement or SSEE utilizes an adjustable half-stop in the condenser front focal plane to occlude a portion of the condenser aperture. A complementary spatial filter located in the objective conjugate rear focal plane attenuates and controls the phase shift of the undiffracted light beam transmitted through the microscope. The degree of attenuation and the phase shift of the background (direct) wave, relative to light scattered and diffracted by the specimen into the aperture region obscured by the half-stop, are adjustable. Images generated by this technique are again similar in appearance to those obtained by differential interference methods, but the single-sideband approach, unlike DIC, does not require polarization optical elements to sandwich the specimen. The system can be adjusted to detect extremely minute phase differences in specimens, as well as anisotropies of refractive indices at high resolution. The technique takes advantage of the fact that only one of the two diffraction sidebands must be captured in order to gain the information regarding the spatial frequency in the specimen that gives rise to that particular diffraction angle. As presented in figure above, single-sideband edge enhancement contrast provides a high-spatial-frequency modulation transfer function superior to that provided by other popular modes of contrast generation.

Extinction

Extinction occurs when one vibration direction of a mineral is parallel with the lower polarizer. As a result no component of the incident light can be resolved into the vibration direction of the upper polarizer, so all the light which passes through the mineral is absorbed at the upper polarizer, and the mineral is black.

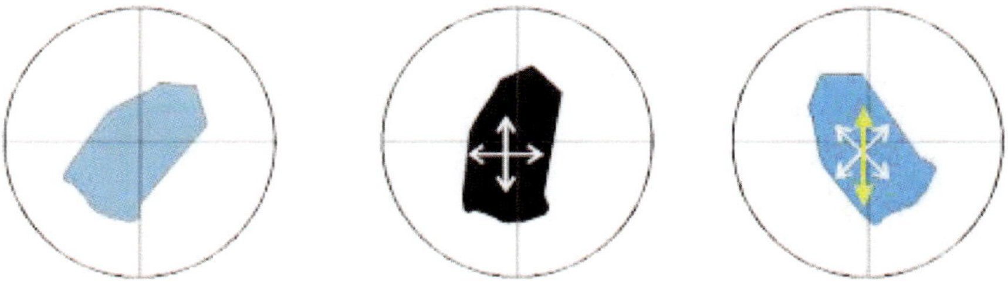

Upon rotating the stage to the 45° position, a maximum component of both the slow and fast ray is available to be resolved into the vibration direction of the upper polarizer. Allowing a maximum amount of light to pass and the mineral appears brightest.

The only change in the interference colours is that they get brighter or dimmer with rotation, the actual colours do not change.

Many minerals generally form elongate grains and have an easily recognizable cleavage direction, e.g. biotite, hornblende, plagioclase.

The extinction angle is the angle between the length or cleavage of a mineral and the minerals vibration directions.

The extinction angles when measured on several grains of the same mineral, in the same thin section, will be variable. The angle varies because of the orientation of the grains. The maximum extinction angle recorded is diagnostic for the mineral.

Types of Extinction

1. Parallel Extinction

The mineral grain is extinct when the cleavage or length is aligned with one of the crosshairs. The extinction angle (EA) = 0°.

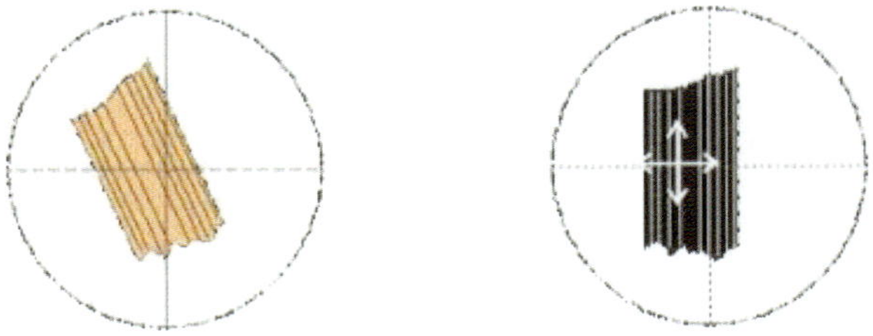

e.g.

- Oorthopyroxene
- Biotite.

2. Inclined Extinction

The mineral is extinct when the cleavage is at an angle to the crosshairs. EA > 0°.

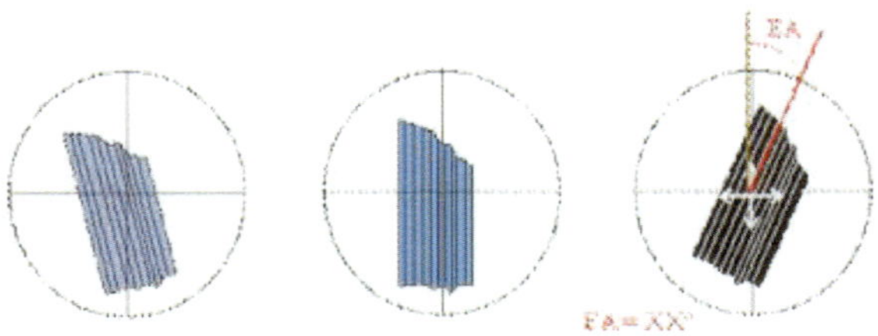

e.g.

- Clinopyroxene
- Hornblende.

3. Symmetrical Extinction

The mineral grain displays two cleavages or two distinct crystal faces. It is possible to measure two extinction angles between each cleavage or face and the vibration directions. If the two angles are equal then Symmetrical extinction exists. $EA_1 = EA_2$.

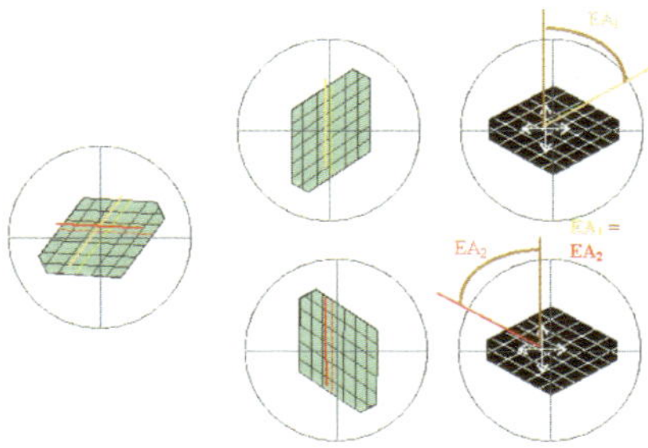

e.g.

- Amphibole
- Calcite.

4. No Cleavage

Minerals which are not elongated or do not exhibit a prominent cleavage will still go extinct every 90° of rotation, but there is no cleavage or elongation direction from which to measure the extinction angle.

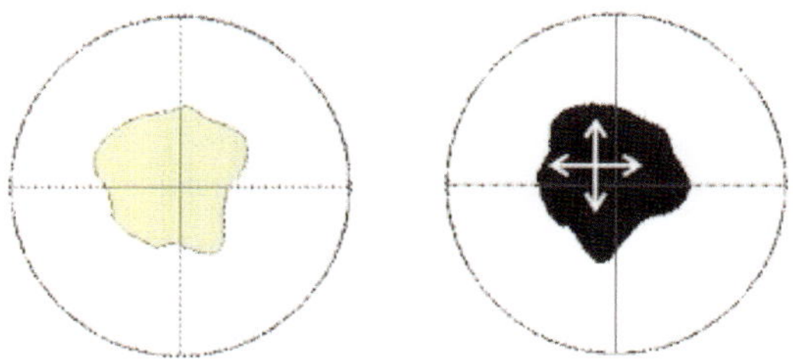

e.g.

- Quartz
- Olivine.

Exceptions to Normal Extinction Patterns

Different portions of the same grain may go extinct at different times, i.e. they have different extinction angles. This may be caused by chemical zonation or strain.

Chemical Zonation

The optical properties of a mineral vary with the chemical composition resulting in varying extinction directions for a mineral. Such minerals are said to be zoned. E.g. plagioclase, olivine.

Strain

During deformation some grains become bent, resulting in different portions of the same grain having different orientations, therefore they go extinct at different times. E.g. quartz, plagioclase.

Petrographic Microscope

A petrographic microscope, also called a *polarizing microscope*, is best described as a compound, transmitted-light microscope to which components have been added to enable the determination of the optical properties of translucent substances. The designation of the microscope as a compound microscope indicates that it has an ocular that focuses on a virtual image of the subject produced in the tube of the microscope by the objective lens.

The professional petrographic microscope has a substage condenser that can be centered and focused. The substage has field and aperture diaphragms. The polarizing components are the upper and lower polarizing devices, the Bertrand lens and its mounting (between the upper polarizing device and the ocular), an accessory flip-in lens for convergent light mounted as the top element of the condenser, and a graduated rotating stage with a removable click stop that can be activated to indicate a 45-degree rotation from any selected direction. The focus knob(s) and the stand are graduated to permit the determination of thickness by differential focusing. Figure below shows a petrographic microscope.

The first polarizing devices were prisms fabricated from crystals of the mineral calcite specially cut and cemented back together according to the plan of Nicol. Such devices were originally called *nicol prisms* . Now they are called *nicols*. Crystals of calcite, $CaCO_3$, were chosen because light that goes through this substance in particular directions is split into two distinctly different rays that are strongly polarized at 90 degrees to each other and that travel at widely different indices of refraction; thus, calcite has a high birefringence. Nicol used the plane on which he cemented the crystals back together as a plane of total reflection for the ray with the index of refraction most different from that of the cement.

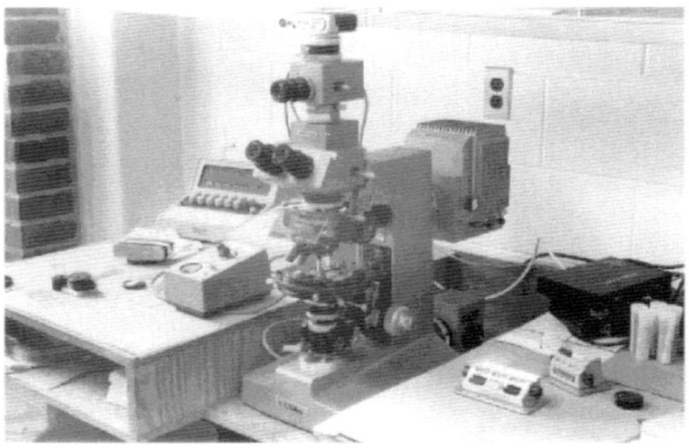

Petrographic microscope

In modern petrographic microscopes, the nicols are polarizing plates fabricated in much the same way as are the lenses in polarized sunglasses. If two polarizing plates of sufficient thickness and quality are superimposed with their polarization directions at right angles to each other (thus,crossed nicols), no light can penetrate the pair. This is because the first polarizing plate excludes all light that is polarized perpendicular to the direction of the plate polarization, concomitantly polarizing the remaining light parallel with the plate polarization, and then the second plate does likewise. Together, all light is excluded. An indication of this effect can be observed by looking through two polarized sunglass lenses superimposed 90 degrees to each other. Polarized sunglasses work because all reflected light is (at least partially) polarized parallel with the substance from which it is reflected and the polarizing material prevents the passage of rays polarized in a horizontal direction, such as that reflected from puddles, snow, automobile surfaces, and pavement. The effect may be observed if you look through a polarized lens at a patch of reflected glare while turning the lens to various orientations relative to the polarization of the glare. (The sky is polarized by reflection from the molecules of the atmosphere.)

A properly adjusted petrographic microscope with nicols of sufficient quality allows no light discernable to the eye to penetrate the upper and lower polarizing devices when their polarization directions are 90 degrees to each other and there is no birefringent substance between them. However, if a birefringent substance (such as a crumpled piece of cellophane) is placed between two polarizing plates that are positioned with their polarization directions 90 degrees to each other, the birefringence of the inner substance polarizes the light that travels through the first polarizing plate in directions parallel to the optical directions of the substance and the optical system will transmit light. The intensity and color of this transmitted light are controlled by the birefringence, optical orientation, and thickness of the interior substance. The lower and upper nicols, or polarizing plates (called the *polarizer and analyzer*, respectively), of the petrographic microscope act as do the plates of polarized material. If both nicols are in the optical path and are oriented at right angles to each other and a birefringent specimen material is on the stage, the amount of the birefringence and many other optical properties may be determined.

In the petrographic microscope, the light is collimated by the condenser into a bundle of beams, all parallel to the optical aids of the microscope. The specimens examined are transparent to translucent thin sections or grain mounts of the material under study. The light beams are polarized in one direction (by the polarizer) before the light reaches the specimen. This light is called *plane polarized light*.

The direction of polarization produced by the polarizer varies from microscope to microscope (usually north-south) and is often adjustable. The analyzer may be placed in the path of the light after it leaves the specimen whenever birefringence or optical directions are being determined.The analyzer is identical in nature to the polarizer and, for most work, the polarization of the analyzer is perpendicular to that of the polarizer (i.e., the analyzer polarization is east-west). Inthe standard orientation, the analyzer allows the transmission of light that is not polarized at rightangles to the light from the polarizer.

When both nicols are in use in this standard orientation, the object on the stage is said to be viewed with crossed nicols or polars. In addition, most petrographic microscopes are equipped with a slot at 45 degrees to the main polarization directions and various retardation plates that can be used in this slot in the determination of a number of optical properties. The most common and most useful of these plates is the 1/4 (one-quarter) wave plate [is correct], or gypsum plate.

Uses of the Petrographic Microscope

The uses of the petrographic microscope include identifying translucent substances by means of their optical properties and by reference to the various charts and tables in the literature. Thus, the composition and identity of these substances and the relationships between the various phases of the material under study may be discovered. From these data, facts concerning the history and the method of formation of the subject material can be deduced. The optical properties of substances given in texts, charts, and graphs have generally been accurately determined by the use of a universal stage on which the substance can be oriented at any desired angle to the optical axis of the microscope and the plane of rotation. In this manner, the optical properties have been determined for various substances. The concrete petrographer does not usually try to attain this degree of accuracy.

Procedures

The identification of aggregate minerals and rocks and concrete reaction products may usually be most efficiently accomplished by knowing which mineral substances are likely; noting the outward physical properties, color, cleavage, and hardness either in a hand specimen or with the stereomicroscope; and using the petrographic microscope to observe the general appearance in polarized light to determine the approximate birefringence, indices of refraction, and some of the other optical properties. The procedures involved include determining some of the following optical properties, listed in order of most common usage:

1. Distinctive cleavage, growth lines, inclusions, and parting patterns.
2. Positions of extinction (directions in the substance where crossed nicols permit the passage of least light).
3. Alignment of the positions of extinction with specific directions (crystallographic axes, cleavage planes, growth lines, etc.) within the subject substance.
4. Birefringence as estimated from the maximum double refraction (highest order of color seen when the positions of extinction are 45 degrees to the nicols) and the thickness of the crystal being examined: The thickness of the subject substance may be determined by differential focusing or by the double refraction exhibited by adjacent substances of known birefringence. The birefringence of the mineral substances is listed in the various charts and tables of mineral properties and is the difference between the highest and lowest index of refraction of the substance. The maximum diffraction is used to determine this property. For this purpose, grain mounts are much more suitable than thin sections. If the substance has a pronounced cleavage or crystal shape that influences the orientation of the particles on the glass slide of a grain mount, some particles should be induced to roll to a new orientation by the nudging of the cover glass with a needle. Birefringence is one of the least variable of the easily determined optical properties. It is relatively constant throughout a family of minerals (such as the feldspar family); the indices of refraction and color may vary from one family member to another.
5. Pleochroic properties (if any) as determined by the change in color and the intensity of the color observed in plane polarized light as translucent substances are rotated through the various positions of extinction (the orientation of these color changes with respect to cleavage and crystallographic axes is an important part of the data on pleochroism).
6. Indices of refraction of a substance determined by comparison in plane polarized light with the index of refraction of the medium (other minerals, mounting epoxy,index of refraction oil, etc.) by which it is surrounded: This is usually done by noting the motion of the Becke line (the bright line that develops as the objective lens is moved out of focus). This determination is aided by the contrast (relief) with which the substance is seen. The greater the difference in the index of refraction between the subject substance and its surroundings, the greater the difference in relief between the substances. This contrast will change as a highly birefringent mineral is rotated from one position of extinction to another and a light ray with a different index of refraction is made parallel with the polarizer. For more exact work, the substance is isolated and fragments are mounted in various standard index of refraction oils.
7. Alignment of the individual indices of refraction of a substance with specific directions (crystal axes, cleavage planes, growth lines, etc.) within the *subject* substance.
8. Number of optical axes, optical sign, and angle between the optical axes: These are determined by means of the conoscopic lens arrangement (i.e., using convergent light and the

Bertrand lens). If the Bertrand lens cannot be focused or is otherwise imperfect, smaller but often sharper optical axis figures may be observed without it by removing the eyepiece lens and looking down the microscope tube at the back lens of the objective. To center the microscopist's eye, a pinhole eyepiece is often used. The data concerning the optical axes and optical sign are often not required in the petrographic study of HCC.

Thin Section

A thin section is a 30 μm (= 0.03 mm) thick slice of rock attached to a glass slide with epoxy. Typical thin section slides are 26 mm x 46 mm, although larger ones can be produced. They are generally covered by another glass slide, a cover slip also attached to the rock with epoxy. The epoxy ideally has an index of refraction of 1.54, although our epoxy is slightly higher, perhaps 1.56.

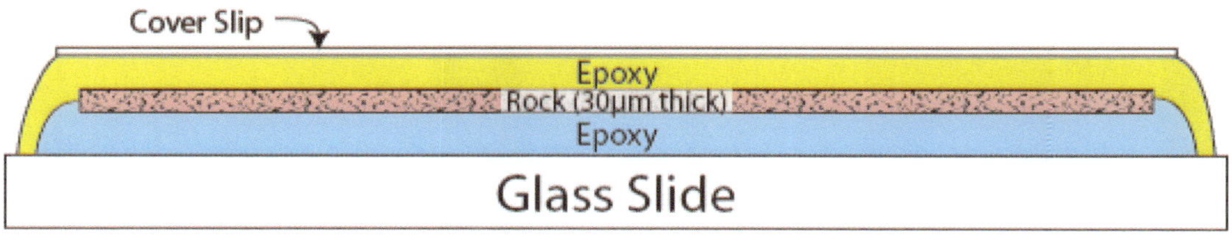

The sections may be left uncovered for chemical analysis on the SEM or electron microprobe. If so, temporary cover slips may be weakly attached with glycerin.

Thin Section Making Equipment

There are five main tools you will use: The slab saw, the trim saw, the grinder, the cut-off saw, and the lap wheels. You will want a Sharpie-type permanent marker, some liquid paper and a fine-tipped permanent marker.

Warnings: Things you can Damage in the Lab

- Be sure when you use either the cut-off saw, or the thin section grinder, that the bottle is less than half full of water. If it gets too full, water can be sucked into the building's vacuum system. This would be bad. Don't do this.
- Be sure to clean and dry the grinding wheels promptly after you finish using each one, or they will rust rapidly. Instructions for this are given below.
- Be sure to use the correct grit for each grinding wheel. From left to right, they get finer and the grit numbers get larger: 120, 240, 400, 600.

Making your Thin Section

Step One: Prepare the Glass Slide

The glass slide you will glue the rock to must be flat in order for the rock section to end up with a constant thickness. In order to achieve this you need to "frost" the slide, which accomplishes two goals:

1. Removes the thick spots on the slide.
2. Adjusts the slide face to be parallel to the grinding wheel's face.

To do this successfully, you need to be able to keep the slide in the same orientation both which you are frosting it and later when you are grinding it. To accomplish this task we use grinding off a corner method, and the slide using the 240 grit grinding wheel.

Here are the steps:

1. Turn on the second wheel from the left.
2. Spray some water from one of the bottles on the wheel until it's completely wet.
3. Sprinkle some abrasive on the wheel - you don't need very much at all.
4. Carefully let a corner of a glass slide drag on the left side of the wheel (if you use the right side, which is moving towards you, it may kick out of your hand).
5. Once a small amount has ground away, you're done. Clean all grit off the slide in the sink.
6. You must clean the wheel promptly, or it will rust. The next section describes how.

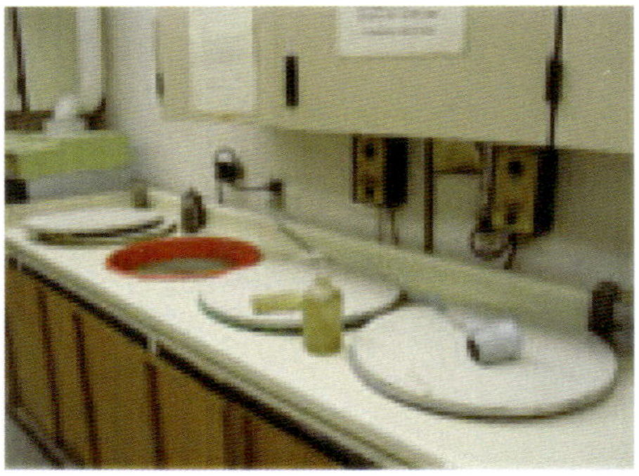

Figure: Grinding wheels. Wheels have particular abrasives: right-most wheel is 120 grit, then 240 grit (open), then 400 grit, then 600 grit. Glass slide corners are ground using the 240 grit wheel.

Figure: Putting water on spinning wheel.

Figure: Putting 240-grit abrasive on spinning, wet wheel.

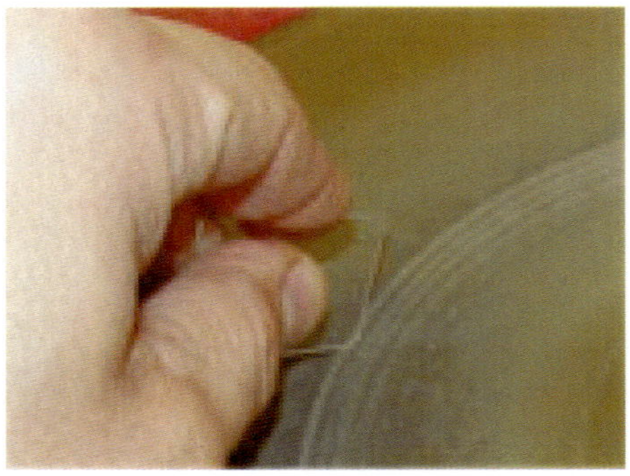

Figure: Grinding the corner of a glass slide on the wheel.

Figure: Glass slide with a corner ground off. Note reflection - it's not yet frosted.

Cleaning the Grinding Wheel

You must clean the grinding wheels promptly, or they will rust. Here's how:

1. Fill a large half-cut jug with water. Clean off a sponge.
2. Pour half of it on the spinning wheel.
3. Clean the spinning wheel with the sponge. Pour some more water on the wheel.
4. Repeat step C until the sponge stops picking up dirt from the wheel.
5. Pour the remaining water around the outside of the plastic bucket to make the grit wash away.
6. Dry the spinning wheel with a paper towel or two.
7. Dry the spinning wheel with the hair dryer until it's totaly dry.
8. Cover the wheel immediately.
9. Clean the bench top that holds the wheels.

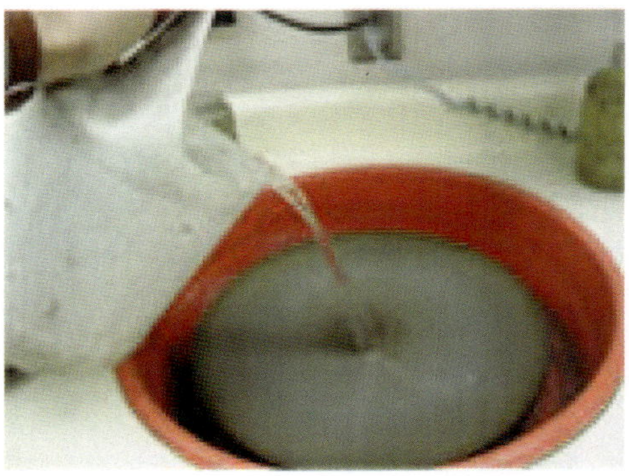

Figure: Pouring water on the grinding wheel for initial cleaning.

Figure: Cleaning the wheel with the sponge until all dirt is gone.

Figure: Inital drying of wheel with paper towel.

Figure: Final drying of wheel with hair dryer.

Step Two: Frost the Glass Slide

You will now "frost" or grind the glass slide to flatten it out and roughen the surface so the epoxy can bind well. To accomplish this, you must always place the slide on the grinder in the same orientation. (I always put the slide on so that the cut-off corner points up and towards me).

1. Turn on the water to the grinder.
2. Find a fairly clean slide (not frosted) and place it on the slide holder of the cut-off saw (wet it if it doesn't stick). This is because the two devices share a vacuum system, and the one not in use must have a slide to create a seal.
3. Wipe clean the slide holder of the grinder. Any grit will produce scratches on the finished section.
4. Place the section on the holder of the grinder.
5. Turn on the vacuum (the valve is between the two instruments; it's on when it points away from the wall).
6. Turn the control towards you until it reads about 70. (This is counter-clockwise; you should see the left part of the grinder move away from the right part which holds the grinding wheel).
7. Turn on the motor.
8. While moving the slide back and forth, slowly turn the control clockwise until the slide begins to contact the grinding wheel.
9. Note the number (on the control) at which contact is made.
10. While still moving the slide back and forth, advance the control wheel gradually about three ticks from the contact value. At this point, grind for a while without advancing the wheel while moving the slide back and forth, to ensure complete and even grinding.
11. Turn off the vacuum and press the vacuum release valve (the black tube from the slide holder connacts to it; it is pressed towards you). When the hissing stops, remove your slide from the holder. (Try not to drag it, or it may get scratched).
12. Dry it off and see if it's all frosted evenly. If not, repeat the above steps until it is. You should not need more than four ticks or so.

13. Once it's all frosted, make a note of the position at which you stopped; you'll need it later. This value is the zero-thickness position, the position at which the grinding wheel contacts the glass slide.

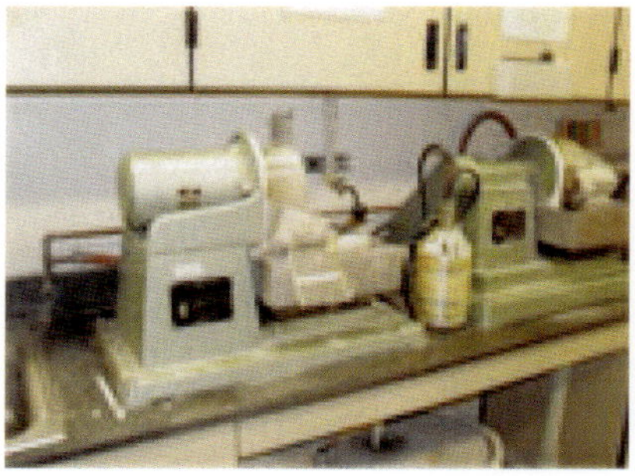

Figure: Cut-off saw (left) and thin section grinder (right). Note that water controls are between sink and cut-off saw, and vacuum controls are between the two.

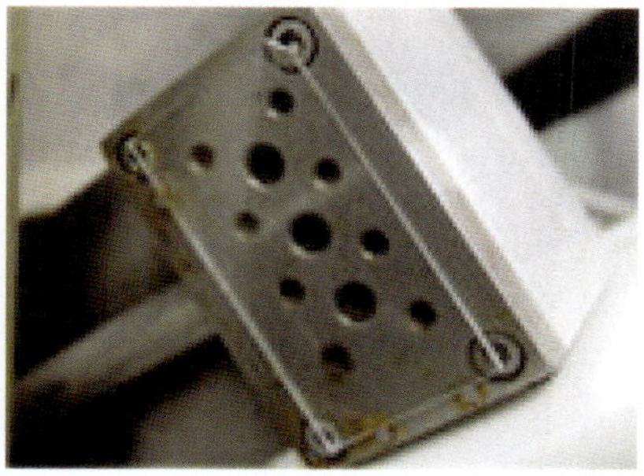

Figure: Slide on grinder showing position including ground corner.

Figure: Placing slide on grinder for frosting. Be sure slide rests against the pins, and place it on the holder with the vacuum turned off.

Figure: Grinder control with numerical gauge.

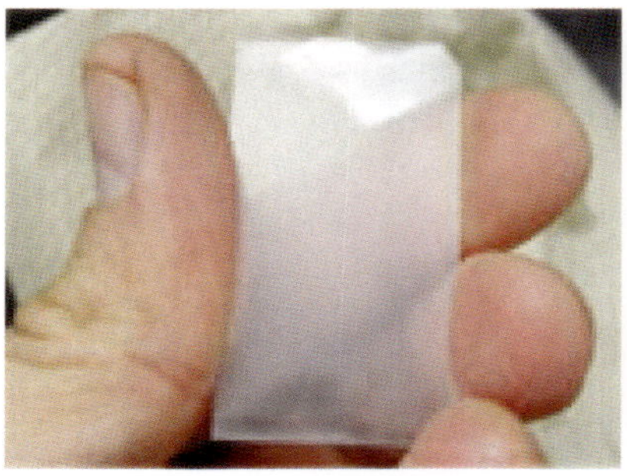

Figure: Slide mostly frosted, but with unfrosted portion at top.

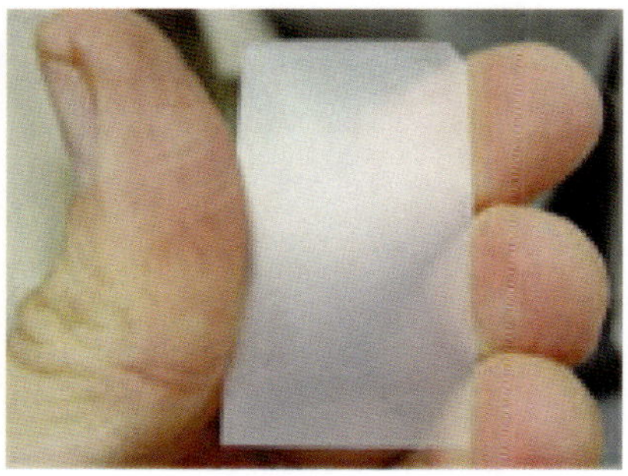

Figure: Slide completely frosted.

Step Three: Mark your Rock

You need to decide where you will cut your rock. This matters, especially with rocks that have a fabric. You generally will want to cut a thin section on a plane perpendicular to any planar fabric, but for particular purposes you may want thin sections in other orientations. Mark a line on the rock with the Sharpie.

Figure: Rock prior to cutting. Pencil for scale. Note that this rock was not marked before cutting, but normally it would be.

Step Four: Cut the Slab

Using the slab saw in the back of the room, you will cut a slab from your rock along the line you marked.

First clamp the sample in the holder so that the line is parallel to the saw blade, sticking out of the holder. You do not make your first cut along this line.

Find crank handle and attach to square protrusion on tray. Wind the tray about 11 turns away from the saw blade. There is some hysteresis, or "slop", so don't count crank turns where the tray isn't moving. Remove the crank handle.

Close the cover, making sure the sides and bottom are tucked inside the lip.

Make the cut. First turn the switch on to get the blade moving, then turn the knob until the weight begins to fall.

Once the cut is done (you can hear it), turn the blade off and stop the weight moving by turning the knob away from you.

Pull the tray back to the beginning, and use the crank to wind the blade back to your line, 11 real turns (where the tray is moving).

Remove the crank handle, and close the cover.

Make the next cut as you did the first. These two cuts should be parallel planes about 8-10 mm apart. The slab should fall onto the catch tray behind the blade.

Step Four: Cut the Slab

Using the slab saw in the back of the room, you will cut a slab from your rock along the line you marked.

1. First clamp the sample in the holder so that the line is parallel to the saw blade, sticking out of the holder. You do not make your first cut along this line!

2. Find crank handle and attach to square protrusion on tray. Wind the tray about 11 turns away from the saw blade. There is some hysteresis, or "slop", so don't count crank turns where the tray isn't moving. Remove the crank handle.
3. Close the cover, making sure the sides and bottom are tucked inside the lip.
4. Make the cut. First turn the switch on to get the blade moving, then turn the knob until the weight begins to fall.
5. Once the cut is done (you can hear it), turn the blade off and stop the weight moving by turning the knob away from you.
6. Pull the tray back to the beginning, and use the crank to wind the blade back to your line, 11 real turns (where the tray is moving).
7. Remove the crank handle, and close the cover.
8. Make the next cut as you did the first. These two cuts should be parallel planes about 8-10 mm apart. The slab should fall onto the catch tray behind the blade.

Figure: Slab saw. Note location for crank in bottom center.

Figure: Slab saw controls. Turn knob to move rock into/through blade, and switch start blade turning (only turn on with cover completely closed!).

Figure: Rock ready for first cut. Note that it sticks out extra far to allow for second cut without removing from clamp.

Figure: After first cut. Next, close knob, pull tray back from saw, then move tray towards blade about 11 turns of the crank handle.

Figure: Ready for second cut.

Figure: After second cut: slab complete (piece lying on catch tray at left).

Clean up the Slab

1. First wash the slab, the leftover rock, and the first cut piece (if there's enough to save) to remove any oil and grit from the slab saw process.
2. Next (while the rocks are drying) set the leftover rock and the first cut piece on a paper towel to dry, but mark the sample number on the towel. It's very easy to mix up specimens at this point, because you may have needed to cut through the number previously marked on the rock. Also, the first cut piece likely has no number. Once these are dry, re-mark them with the sample number. I use liquid paper and an india ink pen, followed by nail polish once it's dry.

Figure: Washing oil and dirt off the rocks after slab cutting.

Figure: Drying speicmens with sample number written on the paper towel.

Step Five: Cut the Chip

You need to reduce the size of the slab to slightly smaller than a thin section. For this you will use the trim saw against the back wall of the room. You need to carefully decide from what part of the slab you want to cut the section.

1. You will use a special rock-cutting blade which is made of metal with diamonds embedded in it. Essentially, such blades are narrow grinding tools. You can even safely put your finger on it as it moves (if it is lubricated with water), however you should not put your fingernail on it.
2. You should wear eye- and ear-protection gear when using this saw.
3. Start the water using the valve above and to the right of the saw, and start the blade using the switch to the right.

4. You just put the slab on the platform and move it into the blade. You can use a glass slide to get the size correct (the chip should be slightly smaller than the slide).
5. After cutting the chip, the side from which the thin section will be cut must be polished to remove marks from the saw blade. For best results you can grind on each of the four wheels, but a useable section will result from just grinding on the seccond (240 grit) wheel.
6. As above, clean the grinding wheel when you are done with it.

Figure: These blades are safe to touch as long as they are lubricated, but don't put your fingernail on it.

Figure: The platform moves, so you just push the slab through the blade.

Figure: Final cut for the chip.

Figure: Grinding the chip will remove any saw marks from the cutting.

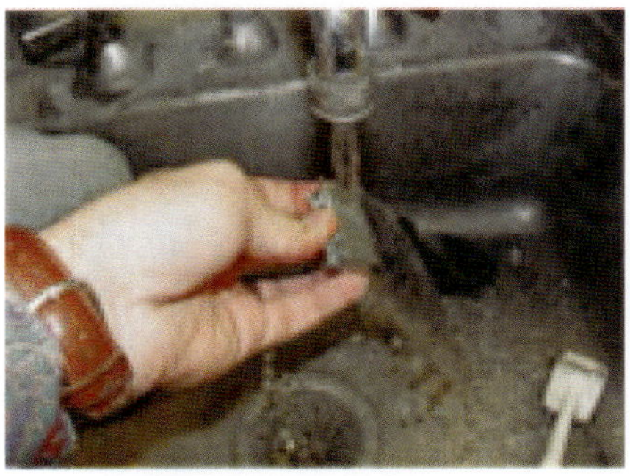

Figure: Washing the chip.

Step Six: Glue the Slide to the Chip

Next, you attach the frosted side of the slide to the side of the chip you just ground down. You need to ensure a constant thickness of epoxy across the section, or when you grind the section down, one end will be thicker than the other.

- Heat up the chip by placing it on one of the hot plates with the polished side up (so it doesn't get dirty). This allows the exopxy to flow more easily and cure faster.
- Write the sample number in pencil on the frosted side of the slide. This number will get covered by epoxy shortly.
- You now need to mix a batch of epoxy.
    - For a single thin section, this will be too much, but it is difficult to accurately mix a small amount.
    - You need to use two parts epoxy (the large bottle) with one part hardener (the small bottle). It's best to use the balance to measure by weight.
    - A small batch would have about 2 grams epoxy and 1 gram of hardener. (Be sure to take into account the weight of the tray!)
    - These materials now must be well-mixed, but while creating as few bubbles as possible. In order to achieve this, I drag the coffee stirrer in a sinuous pattern first in one direction, then in the same pattern rotated 90 degrees.
- Once the epoxy is well-mixed, spread a few drops across the top (polished side) of the warmed chip.
- Many rocks are porous (even some that you might not expect) and you must let the epoxy penetrate fully, or the thin section will not survive. So, spread epoxy, then wait for it to soak in (5 minutes) then spread more epoxy. Repeat until no more epoxy soaks into the chip.
- Place the slide, frosted side down on the epoxy. It's best to put one side down, then let the other side fall to avoid trapping bubbles.
- Move the slide around with your finger or a pencil eraser. This will squeeze out extra epoxy to achieve a constant thickness, and if you work at it, you can remove most bubbles this way as well. While doing this, be sure that the sample number gets coated with epoxy.
- Let this sit to cure. Check it periodically for the first 5-10 minutes to be sure the slide has not slid off the chip.
- The epoxy will cure in about 20-30 minutes. Do not take the next step before that.

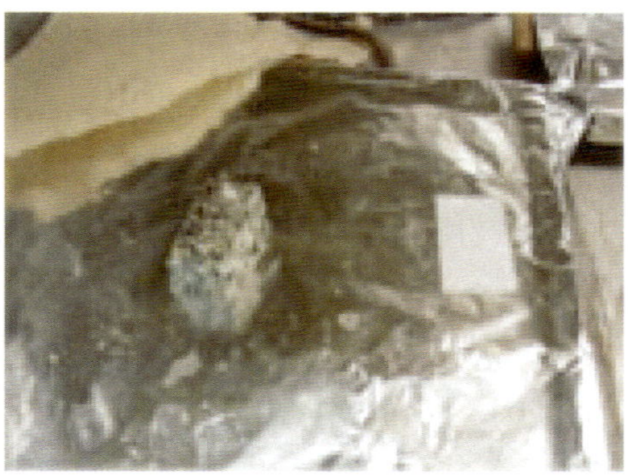

Figure: Cut and polished chip warming up on the hot plate..

Figure: Sample number marked in pencil on frosted side of slide.

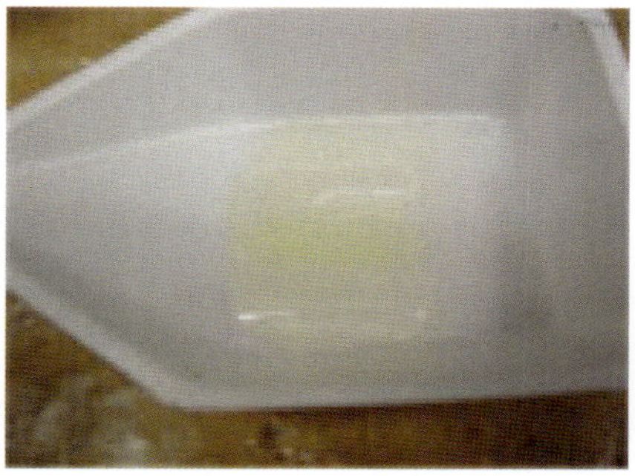

Figure: Epoxy and hardener prior to mixing.

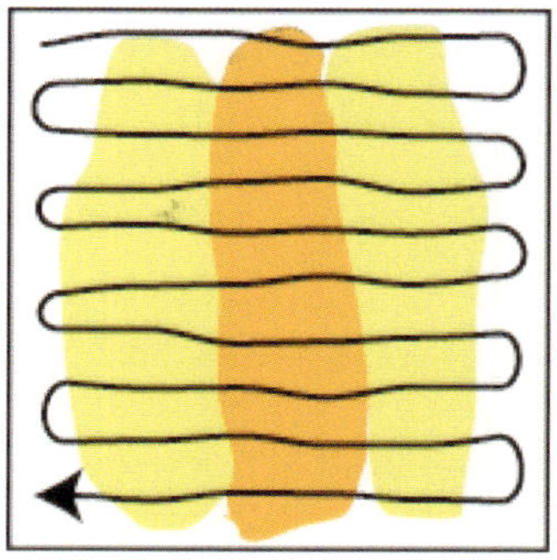

Figure: Pattern for mixing. Should be rotated 90 degrees after each time through pattern.

Figure: Spreading epoxy on chip.

Figure: Chip coated with epoxy after allowing it to penetrate.

Figure: Placing slide on chip, frosted side down.

Figure: Moving slide around to expel bubbles and extra epoxy, and coat sample number.

Step Seven: Cut Off the Chip From the Slide

Now you have a rock chip epoxied to a glass slide. You will next cut most of the chip off, leaving a thin slice attached.

1. Place a blank glass slide on the grinder to block the vacuum.
2. Turn on the water to the cut-off saw.
3. Place the slide on the cut-off saw (note direction of ground corner) and turn on vacuum to hold it. Be sure edges of slide are lodged against pins in holder.

4. Turn on saw motor.
5. Use the handle to move the chip into the blade. Go very slowly, or you can break your slide and have to start over.
6. Once the chip is cut off the slide, retrieve the chip from the water tray and set it aside (it will likely need to be labeled with the sample number).
7. Turn off the vacuum, and saw motor.
8. Using the button (as with the grinder), break the vacuum. Once the hissing stops, remove your slide from the saw.
9. Rinse your slide to remove any particles.

Figure: Slide with chip mounted in cut-off saw.

Figure: Slowly cutting off the chip from the slide.

Figure: Chip nearly cut from slide.

Figure: Chip and slide separated.

Step Eight: Grind your Slide to the Correct Thickness

You must now grind away much of the rock that remains on your slide, but not all of it. This is the step in which most thin sections go bad. The key is: go slow, especially near the end.

- Place a blank glass slide on the cut-off saw to block the vacuum.
- Turn on the water to the grinder.
- Turn the control towards you (counter-clockwise) a full rotation to zero.
- Be sure the slide holder and the slide are clean and free of any grit or particles.

- Place the slide on the grinder (maintaining the same orientation using the corner you notched).
- Turn on the vacuum and grinder motor.
- Moving the slide back and forth with the handle, gradually advance the control until the slide contacts the grinding wheel.
- While moving slide back and forth across the grinding wheel with the black handle, gradually advance the control until you reach a point about 15 ticks below your "zero value".
- Remove the slide (turn off the vacuum, break the vacuum with the button, then remove the slide).
  - If you can, then proceed slowly until those minerals achieve the correct maximum interference color (e.g., quartz should show a maximum interference color of a pale straw yellow)
  - If you cannot, then advance one tick at a time until you can identify a mineral. If you are within 5 ticks of your zero value, and you still can't identify anything, stop and ask someone who knows more than you.
  - Go slow, especially when you are close to the correct thickness! It's quite easy to go from slightly too thick to slightly too thin (or ground completely away), *even without advancing the position control at all!*
- Troubleshooting
  - Warning: Most sections fail at this stage because you are grinding too fast! Remember, any time the grinder makes noise against a slide, you are removing rock. You often don't have to advance the wheel at all to just grind a small thickness away.
  - If your section comes out with the minerals all cracked: you were grinding too fast.
  - If your section comes out with the edges thin and the center thick: you were grinding too fast.
  - If your section comes out thick on one side and thin on the other, you either didn't frost the slide well, or there was an uneven epoxy layer.

Figure: Unground slide mounted in grinder holder.

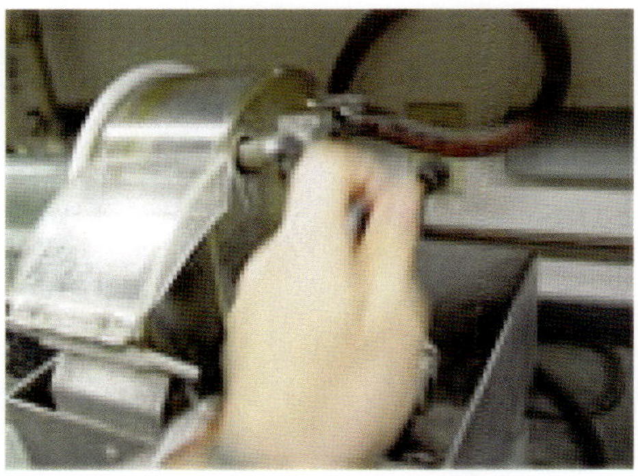

Figure: Moving slide across grinding wheel while advancing control (with right hand).

Figure: Checking the slide thickness using the microscope.

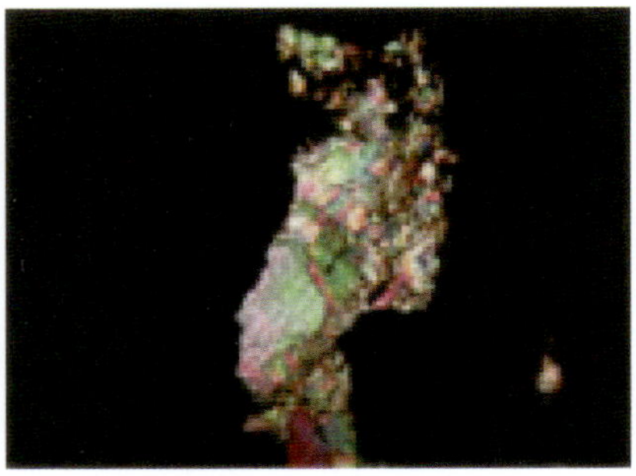

Figure: Quartz in a thin section, way too thick.

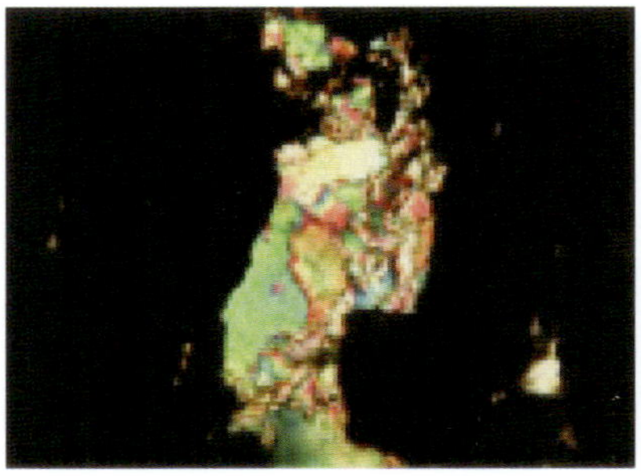

Figure: Quartz in a thin section, still fairly thick.

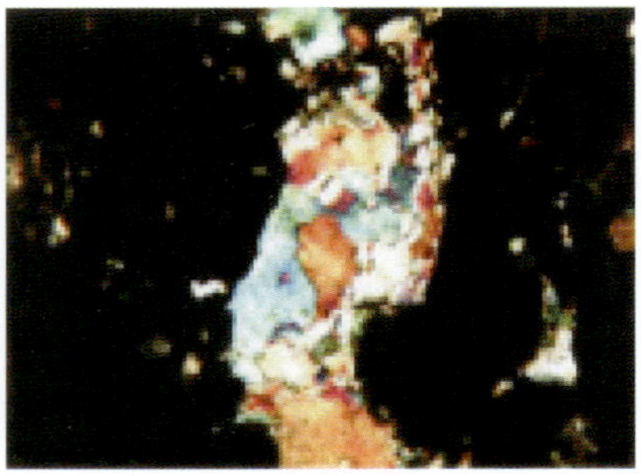

Figure: Quartz in a thin section, still a bit thick.

Figure: Quartz in a thin section, now too thin. At this point, much of the remaining section has been ground away.

Step Nine (Version 1): Add A Cover Slip

If you do not expect to perform chemical analysis on your minerals in this thin section, you should add a cover slip to protect the section from damage, and increase the clarity observed in the microscope. This process is a simple one:

1. Make sure the section is clean and free of grit or dirt.
2. Place it on the hot plate.
3. Mix up a small batch of epoxy and hardener.
4. Place a small drop of epoxy on the section.
5. Drop a cover slip on the drop.

6. Move it around to expel bubbles and fully coat the section.
7. Let it cure.
8. After it has cured, there may be extra epoxy on the top, sides, or bottom. You can remove this, *very carefully*, with a razor blade.

Figure: Putting a drop of epoxy on the ground slide.

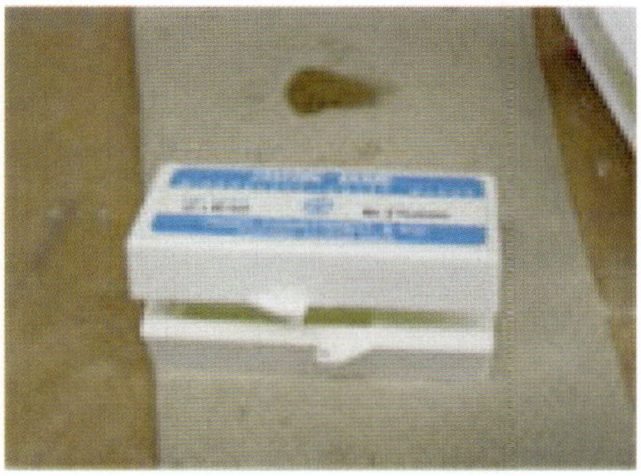

Figure: Container for cover slips.

Figure: Dropping a cover slip on the slide. Put one end down first in order to prevent bubbles.

Figure: Moving cover slip to spread epoxy and expel bubbles.

Step Nine: Polish for Electron Microprobe/SEM-EDS Analysis

If you plan to perform chemical analysis on your minerals, then you need to polish the section much better than the grinder polishes it. This is because these analysis tools require a flat, smooth surface at the micrometer scale. You perform this polishing step using different tools in a different room.

The key diffference for our purposes here is that you should stop grinding when the section is still a little bit thick, because you will be removing material during the polishing process.

The polishing apparatus is located on the ground floor, at the other end of the building (near the XRD).

Step Ten: Clean up the Place

You must clean the lab thoroughly when you are done using it. You may store samples in progress in one or more trays clearly labeled with your name and the date. These trays are placed in the lower cabinets in the thin section lab room, or (for grad students) in the cabinets in the center of the adjacent room.

1. Make sure that the grading wheels are clean and dry.
2. Completely dry off the dark metal section holders on both the cut-off saw and the grinder.
3. For each of those two tools, spread some oil on a paper towel, and coat the dark metal portion thoroughly with oil, then place the oiled paper towel underneath the section holder.
4. Empty out the water from the vacuum bottle (covered with tape).
5. Place any epoxy stirrers crosswise on the epoxy holder so that when the epoxy dries, both can be reused.
6. Mark your name in the log sheet.
7. Turn off the lights and close the door when you leave.

Figure: Drying off the section holder.

Figure: Oiling the paper towel.

Figure: Oiling the section holder.

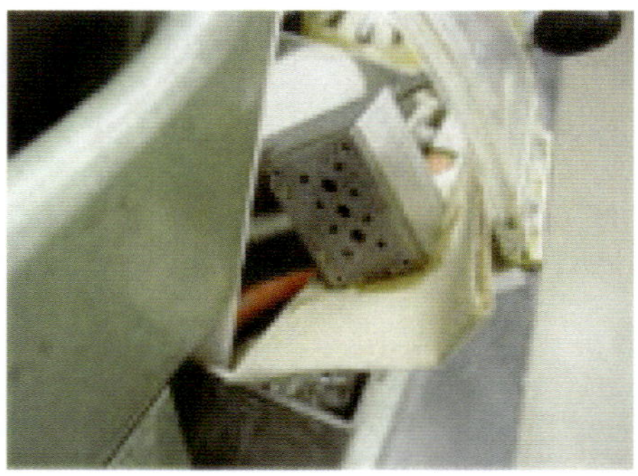

Figure: Tool dried, oiled, and paper towel placed for storage.

Figure: Removing tubing from vacuum bottle.

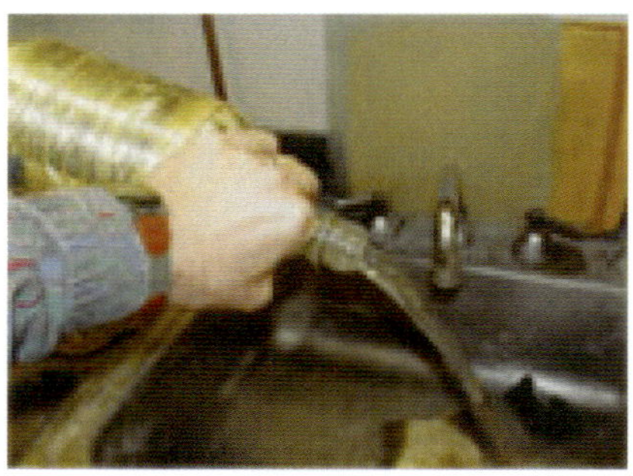

Figure: Emptying the vacuum water bottle.

Figure: Epoxy stirrer and tray stored for curing and reuse.

Figure: Filling out the log sheet.

References

- Optical-mineralogy: kurdistan-geology.com, Retrieved 26 April 2018
- Optical, property: minerals.net, Retrieved 16 May 2018
- Obliqueintro, oblique, microscoperesource, techniques: olympuslifescience.com, Retrieved 16 March 2018
- Optical, earthsciences: brocku.ca, Retrieved 26 July 2018
- Thinsections: davehirsch.com, Retrieved 16 July 2018

Chapter 4

Magnetic Mineralogy

Magnetic mineralogy is the science concerned with the study of the magnetic properties of minerals. How a mineral contributes to the total magnetism of a rock is dependent on the kind of magnetic order or disorder. This chapter closely examines the key concepts of magnetic mineralogy with topics such as magnetite, greigite, hematite, kamacite, etc. for an extensive understanding of the subject.

Magnetic mineralogy (or magnetomineralogy) deals with the type, concentration and grain size of magnetic minerals with an aim to apply such information to magnetic prospecting, paleoclimatology, monitoring of environmental pollution or providing a basis for interpretation of paleomagnetic results where a magnetization stable over a long time span is highly desired. Due to the sensitivity of the state-of-the-art instruments, the magnetic measurements are nowadays applicable to virtually all materials, i.e. rocks, soils or even organic materials, including those whose magnetism is so weak that they are commonly considered as non-magnetic.

We can measure two basic but invaluable magnetic properties: low-field magnetic susceptibility and remanent magnetization.

Case Study

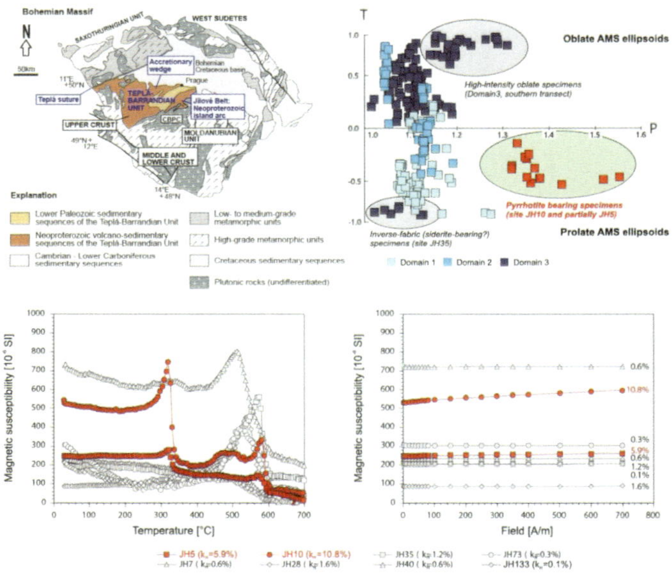

The identification of pyrrhotite as the carrier of prolate, high-degree magnetic fabric in the flysch-like sediments of the Teplá-Barrandian Unit, Bohemian Massif.

When investigating the magnetic fabric in the Neoproterozoic graywackes and siltstones, several different fabric types of were identified based on its magnitude and symmetry. Besides a commonly-observed, phyllosilicate-carried fabric characteristic for deformed sedimentary rocks (medium-degree, oblate or prolate), both high-degree oblate and prolate fabrics was observed. While the oblate fabric can be artibuted to the zones of higher deformatrion, the high-degree prolate fabric (highlighted in red) is rather difficult to be achieved in rocks where the fabric is solely carried by the oblate-fabric phyllosilicate grains. Using an AGICO Kappa bridge coupled with a CS-4 temperature control unit, magnetic susceptibility was measured as a function of temperature and amplitude of applied AC field. Both rock magnetic methods independently identified the presence of pyrrhotite based on its Curie point and a pronounced increase of susceptibility as a function of field, respectively. Resulting from this simple test, pyrrhotite-bearing specimens must be excluded from the entire sample set when magnetic fabric is interpreted in terms of strain variations in the studied geological unit.

Magnetic susceptibility very sensitively reflects any slight variations in types and concentration of magnetic minerals further exceeding the capability of most conventional analytical methods. Its analytical power can be further extended when it is measured as a function of temperature, amplitude of the applied field, or at various operating frequencies.

In addition to magnetic susceptibility, we can measure the artificial remanent magnetization, either isothermal (IRM) or anhysteretic (ARM), laboratory imparted in various fields, usually much stronger than the Earth's magnetic field.

Magnetite

Magnetite is a ferromagnetic mineral with the chemical formula Fe_{3O4} and the common chemical name ferrous-ferric oxide, which indicates the mineral comprises both a ferrous component, FeO (wüstite), and a ferric component, Fe_2O_3 (hematite). Magnetite is one of several types of iron oxide and its official (IUPAC) name is iron(II,III) oxide. It is a member of the spinel group of minerals, which crystallize in cubic and octahedral patterns, and its crystals are black and opaque. The most magnetic of all naturally occurring minerals on Earth, magnetite occurs in some places as naturally magnetized stone called lodestone and was used as an early form of magnetic compass. Magnetite dissolves slowly in hydrochloric acid.

Magnetite mineral is valuable as an iron-bearing ore. In addition, as it carries the dominant magnetic signature in rocks and it tends to lock in the magnetic pattern it carried as it was last hardening, magnetite has played a critical role in understanding plate tectonics. Changes in the oxygen content of the Earth's atmosphere can be inferred by studying sedimentary rockscontaining magnetite. Moreover, interactions between magnetite and other oxide minerals have been studied to determine the oxidizing conditions and evolution of magmas over geological history.

Small grains of magnetite occur in almost all igneous rocks and metamorphic rocks. Magnetite also occurs in many sedimentary rocks, including banded iron formations. In many igneous rocks, magnetite-rich and ilmenite(a titanium iron oxide)-rich grains occur that precipitated together from magma.

Properties

The most striking property of magnetite is very strong ferrimagnetism. It makes the mineral easily identifiable because it is strongly attracted to a hand magnet. Ferrimagnetism is caused by opposing, but unequal magnetic moments within the crystals which results in permanent and spontaneous magnetization of the material.

The presence of di- and trivalent iron in the crystal lattice is the reason why magnetite is so strongly magnetic. Divalent (+2) and trivalent (+3) iron have unequal magnetic moments that are not balancing each other. Magnetite is the most magnetic mineral.

High iron content gives magnetite its opaqueness and black color. Spinel which shares the same structure is variably colored and transparent because it contains magnesium and aluminum instead or iron.

Magnetite is dense (specific gravity 5.20) mineral. This is considerably above common silicate minerals (usually 2.5–3.5) which is why rocks containing appreciable amount of magnetite feel heavy in hand sample. Hardness is about 6 on the Mohs scale. Magnetite has no cleavage but parting may be distinct. Crystals are brittle and fracture is uneven.

This is how magnetitic sand aligns itself in the presence of a strong external magnetic field. There is a neodymium magnet placed beneath the sample. Crystals from Talofofo Beach, Guam, USA. Width of view 10 mm.

Occurrence

Magnetite is a very common (but usually accessory) mineral in igneous and metamorphic rocks. It occurs in a wide variety of igneous rocks as small octahedral or anhedral grains. It may form larger segregations in contact-metasomatized carbonate rocks (skarns) where it is associated with calcite and calc-silicate minerals like diopside, andradite, actinolite, tremolite, etc.

Massive variety may also occur in some mafic layered intrusions. It may form in regionally metamorphosed rocks where it forms at the expense of iron hydroxides (goethite, limonite) and oxides (hematite).

It is the main iron-bearing mineral in the oldest Algoma-type banded iron formationswhere it is associated with chert.

Magnetite is among the most common minerals in heavy mineral fraction of sand. Its grains in sand are generally much smaller than lighter mineral grains because of different settling velocity. Most magnetite grains in sand are rounded but some show characteristic octahedral morphology. It is never elongated because of cubic (isometric) crystal system.

Magnetite is common in sand because it is abundant in many rock types and it is also moderately resistant to weathering. In some places beach sand may be so concentrated in magnetite that it could be used used as an iron ore. In New Zealand a sand deposit called Ironsand is used to make steel.

Magnetite is altered in the weathering environment to hematite, goethite or other iron oxides and hydroxides. Martite is a pseudomorphous hematite after magnetite.

Magnetite is a common heavy mineral in sand. This rock is a metamorphosed sand deposit which seems to be very rich in heavy minerals magnetite (black) and garnet (red). Varanger Peninsula, Northern Norway. Width of sample 18 cm.

Actinolite (green) with magnetite and calcite. Kiruna, Sweden. Width of sample 8 cm.

Magnetite crystals forming black stripes in light-colored sand. It is one of the most common constituents of heavy minerals in sand. White Park Bay, Northern Ireland.

Magnetite with amphibole group mineral tremolite in skarn.

Skarn is a contact-metasomatic rock. It forms when hot silicic magma comes to contact with carbonate country rocks (dolomite, limestone, marble). The result is unusual assemblage of calc-silicate minerals like tremolite, diopside, andradite, wollastonite, etc. These rocks also frequently contain ore minerals because late-magmatic fluids are usually enriched in incompatible chemical elements that have no place in the crystal structure of common magmatic minerals. Skarn was originally a miners term for the gangue minerals (calc-silicates) surrounding the ore veins. Width of sample 8 cm.

Magnetite in skarn. Gangue minerals are serpentine and talc. These minerals hint that there must be a major source of magnesium. These rocks indeed formed when magma intruded and reacted with dolomitic (Mg-Ca-carbonate) marble. Tapuli, Sweden. Width of sample 11 cm.

Skarn sample with magnetite, diopside (Ca-Mg-pyroxene), and calcite.
Tapuli, Sweden. Width of sample 12 cm.

Magnetite is a common hydrothermal mineral that occurs in quartz veins with other ore minerals. This sample also contains quartz (white), pyrite, and chalcopyrite. Hannukainen, Finland. Width of sample 11 cm.

Uses

Magnetite is a major source of iron. Banded iron formations are precambrian metasedimentary rocks where the iron-bearing phase is usually either magnetite or hematite. Very rich magnetitic iron ore is in Kiruna (northern Sweden) although the formation details are not clear (it is not banded iron formation). Skarn-related iron ores are also mined although they tend to be less voluminous. Iron may be also extracted from placer deposits (heavy mineral sand).

It is industrially used as a feedstock in the manufacture of other iron-bearing materials. Magnetite has been used to make high density concrete to nuclear reactors. It is also used as a black pigment.

Naturally magnetized magnetite is called lodestone. Normally it is only attracted to hand magnet, but magnetite itself does not attract objects mader of iron. Lodestone is different because it does that too and it readily aligns itself along the magnetic lines of the Earth. This makes lodestone useful in navigation as a natural magnetic compass. It is not entirely clear why some magnetites are naturally magnetized, but lodestones contain inclusions of maghemite (spinel group mineral) and one theory associates it with magnetic fields surrounding lightning bolts. This could explain why lodestones are found close to the surface, not from deep iron mines.

Magnetite crystals have been found in the brains of several species, including humans. It has been hypothesized that birds could make use of it to navigate, but it is not clear what benefits can they provide to humans.

Massive chunk of iron ore which is composed of almost pure magnetite. Iron ore from Kiruna is world-famous as a very rich high-grade ore. The sample feels very heavy when compared to usual silicate rocks. Kiruna, Sweden. Width of sample 13 cm.

Magnetite with jasper and hematite. These minerals come from the hydrothermally altered oceanic crust. Hot newly formed oceanic crust at the mid-ocean ridge is full of cracks which allow seawater to intrude the crust. Water heats up when circulating within the rocks and leaches metals out of the basaltic crust. Metals are precipitated when this very hot and metal-rich water enters the ocean again through black smokers. These metal deposits are known as SedEx-type (sedimentary exhalative) ore deposits. Løkken ophiolite, Norway. Width of sample 13 cm.

Algoma-type banded iron formation (BIF) from the Archaean. Magnetite is the principal iron-bearing ore mineral in these very old iron ores. Banded iron formation is the main source of iron although the majority of these deposits are from the Proterozoic. Bjørnevatn, Norway. Width of sample 17 cm.

Superior-type banded iron formation from Kryvyi Rih, Ukraine. Superior-type BIFs are the main source of iron. Iron-bearing mineral in these rocks is usually either hematite or magnetite. Width of sample 10 cm.

Magnetite in quartz. Bjørnevatn, Norway. The original banding of BIF is disturbed by the metamorphic processes. Width of sample 11 cm.

Iron ore from Kiruna. The main minerals are magnetite, calcite, actinolite, and apatite. Kiruna is the largest iron mine in Europe. Yet the formation details of these rocks are still poorly understood. Width of sample 14 cm.

Magnetite with feldspar. Kiruna, Sweden. Width of sample 16 cm.

Magnetite with calcite (white) and pyrite (iron sulfide). Kiruna, Sweden. Width of sample 14 cm.

Magnetite in syenite porphyry. Kiruna, Sweden. Width of sample 15 cm.

This is sand from the North Island of New Zealand. It is used as an iron ore. The black grains are titanomagnetite (total titanium content of the sample is 4 percent). Iron makes up 20 percent of the sample (XRF data). Yellow grains are silicate minerals. Width of view 10 mm.

It is a major constituent in a heavy mineral fraction of sand. Lots of black minerals on this gold pan are magnetite grains. There is also gold (yellow spots). Tankavaara, Finland.

Maghemite

Maghemite (Fe_2O_3, ?-Fe_2O_3) is a member of the family of iron oxides. It has the same structure as magnetite, that is, it is spinel ferrite and is also ferrimagnetic.Maghemite can be considered as an Fe(II)-deficient magnetite with formula where represents a vacancy, A indicates tetrahedral positioning and B octahedral.

a) (Fe_2O_3, ?-Fe_2O_3)

b) Maghemite Chemical Formula: Gamma-Fe+++$_2O_3$

c) Maghemite Empirical Formula: $Fe_3+_2O_3$

Maghemite was named in 1927. It is blue with a grey shade, white, or brown. It has isometric crystals. Maghemite is formed by the topotactic oxidation of magnetite.

Synonyms

maghemite, ferrimagnetic, synthetic maghemite, ferrimagnetic ferric oxide, oxymagnite, sosmanite, spinel ferrite, magnetic maghemite nanoparticles, iron oxide, magnetite, black iron oxide, CAS# 1309-37-1.

Chemical Properties

In some cases, various grades are available between 98% and 99.9%.

Physical Properties

a) Various granule and powder granulations down to nanoparticles

b) Targets

c) Sintered pieces

Typical Applications

It is a ferro magnetic oxide, has a spinel structure and is commonly used in magnetic tape. For example in the magnetic layer of audio tape, floppy disk, digital tape etc. The particles can be magnetised to represent data. MICR (Magnetic Ink Character Recognition) also uses iron (III, II) oxide compounds, suspended in an ink which can be read by special scanning hardware.

Packaging

50 lbs. bags, poly-lined fiber drums, bulk bags or to customer specifications.

Specifications

Molecular Weight (g/mol.)	159.69
Apparent Density (g/cm3)	4.9
System	Isometric
RTECS Number	NO7400000
Boiling Point (°C)	
Magnetism	Strongly magnetic
Radioactivity:	Not radioactive
Mohs Hardness @20° C	6
Transparency	Opaque
Color	Brown, brick-red
Morphology:	Massive
Crystallography	Tetragonal

Greigite

Greigite (Fe_3S_4) is a ferrimagnetic mineral with vital functions in both the bio-geochemical cycle and novel technological applications.

It is probably formed by bacteriological reduction of iron and is widespread in nature including lacustrine sediments, magnetotactic bacterium and hydrothermal vein deposits. Greigite has attracted great interests in geophysics and biology because it records the ancient geomagnetic field and environmental processes that are important for paleomagnetic and environmental magnetic studies.

Physical Properties: Hardness = 4–4.5 VHN = 401–423 (50 g load). D(meas.) = 4.049 D(calc.) = 4.079 Strongly magnetic.

Optical Properties: Opaque. Color: Pale pink, tarnishing to a metallic blue; sooty black when very fine grained; in polished section, pale creamy white. Luster: Metallic.

R: (400) 30.5, (420) 31.1, (440) 31.7, (460) 32.2, (480) 32.7, (500) 33.2, (520) 33.6, (540) 34.2, (560) 34.7, (580) 35.3, (600) 35.9, (620) 36.4, (640) 37.1, (660) 37.6, (680) 38.1, (700) 38.7

Cell Data: Space Group: F d3m. $a = 9.876$ $Z = 8$

X-ray Powder Pattern: Kramer-Four Corners area, California, USA.

2.980 (100), 1.746 (77), 2.470 (55), 3.498 (32), 1.008 (31), 1.901 (29), 1.105 (16)

Chemistry

	(1)	(2)	(3)
Fe	56.5	55.5	56.64
Cu	0.08	0.2	
Ni	0.10		
Zn	0.01		
Mn		0.1	
Cr	0.14		
Sb		1.3	
AS	0.38		
S	42.2	42.2	43.36
Total	99.41		100.00

(1) Zacatecas, Mexico; corresponds to $(Fe_{3.08}As_{0.02}Cr_{0.01}Ni_{0.01})_{\Sigma=3.12}S_{4.00}$.

(2) Cornwall, England; by electron microprobe, corresponds to $(Fe_{3.08}As_{0.02}Cr_{0.01}Ni_{0.01})_{\Sigma=3.12}S_{4.00}$. $(Fe_{3.04}Sb_{0.03}Mn_{0.01}C_{0.01})\Sigma =_{3.09} S_{4.00}$.

(3) $Fe^{2+} + Fe_2^{3+} S_4$.

Mineral Group: Linnaeite group.

Occurrence: Formed in lacustrine beds consisting of interbedded calcareous clays, silts, and fine- to medium-grained arkosic sands; in varvelike laminae of grayish black sulfide-bearing clays and fine silts; possibly as the result of seasonal variation in the composition and physical condition of lake waters; also in hydrothermal vein deposits; formed in magnetotactic bacteria, and by sulfate-reducing bacteria.

Hematite

Hematite is one of the most abundant minerals on Earth's surface and in the shallow crust. It is an iron oxide with a chemical composition of Fe_2O_3. It is a common rock-forming mineral found in sedimentary, metamorphic, and igneous rocks at locations throughout the world.

Oolitic Hematite: A specimen of oolitic hematite iron ore. Oolites are tiny round spheres of chemically precipitated hematite. The specimen in the figure is about four inches (ten centimeters) across, and the largest oolites are a few millimeters in diameter.

Hematite is the most important ore of iron. Although it was once mined at thousands of locations around the world, today almost all of the production comes from a few dozen large deposits where significant equipment investments allow companies to efficiently mine and process the ore. Most ore is now produced in China, Australia, Brazil, India, Russia, Ukraine, South Africa, Canada, Venezuela, and the United States.

Hematite has a wide variety of other uses, but their economic significance is very small compared to the importance of iron ore. The mineral is used to produce pigments, preparations for heavy media separation, radiation shielding, ballast, and many other products.

Hematite's Streak: All specimens of hematite will produce a reddish streak. The streak of a mineral is its color in powdered form when scraped across a streak plate (a small piece of unglazed porcelain used to produce a small amount of mineral powder). Some specimens of hematite will produce a brilliant red streak, others will produce a reddish brown streak. Care is needed when testing a specimen of hematite with a metallic luster. These specimens are often brittle and leave a trail of debris along with the streak. That debris is not a powder - it is a trail of fragments. So, to assess the streak, the loose particles must be gently shaken free from the streak plate or very lightly brushed off. This leaves behind the powder that is embedded within the textured surface of the streak plate. In the figure above, the streak on the left has been cleaned of fragments, and you can see that it is a reddish brown. The streak on the right still has a trail of glittery fragments that must be gently removed for proper evaluation.

Physical Properties of Hematite	
Chemical Classification	Oxide
Color	Black to steel-gray to silver; red to reddish brown to black
Streak	Red to reddish brown
Luster	Metallic, submetallic, earthy
Diaphaneity	Opaque
Cleavage	None
Mohs Hardness	5 to 6.5
Specific Gravity	5.0 to 5.3
Diagnostic Properties	Red streak, specific gravity
Chemical Composition	Fe_2O_3
Crystal System	Trigonal
Uses	The most important ore of iron. Pigment, heavy media separation, radiation shielding, ballast, polishing compounds, a minor gemstone

Physical Properties of Hematite

Hematite has an extremely variable appearance. Its luster can range from earthy to submetallic to metallic. Its color ranges include red to brown and black to gray to silver. It occurs in many forms that include micaceous, massive, crystalline, botryoidal, fibrous, oolitic, and others.

Even though hematite has a highly variable appearance, it always produces a reddish streak. Students in introductory geology courses are usually surprised to see a silver-colored mineral produce a reddish streak. They quickly learn that the reddish streak is the most important clue for identifying hematite.

Hematite is not magnetic and should not respond to a common magnet. However, many specimens of hematite contain enough magnetite that they are attracted to a common magnet. This can lead to an incorrect assumption that the specimen is magnetite or the weakly magnetic pyrrhotite. The investigator must check other properties to make a proper identification.

If the investigator checks the streak, a reddish streak will rule out identification as magnetite or pyrrhotite. Instead, if the specimen is magnetic and has a reddish streak, it is most likely a combination of hematite and magnetite.

Specular Hematite: Specular hematite, sometimes called "micaceous hematite," has a metallic luster and appears to be a rock composed of shiny mica flakes. Instead those flakes are hematite.

Even though this hematite has a silver color, it still produces a reddish streak - which is a key to hematite's identification. Hardness testing on specular hematite is difficult because the specimens tend to crumble. This specimen is about four inches across (ten centimeters) and was collected near Republic, Michigan.

Banded Iron Formation: Close-up of a banded iron formation. In this specimen, bands of hematite (silver) alternate with bands of jasper (red). The rock produced where these formations are mined is often called "taconite." This figure spans an area of rock about one foot (30 centimeters) wide.

Composition of Hematite

Pure hematite has a composition of about 70% iron and 30% oxygen by weight. Like most natural materials, it is rarely found with that pure composition. This is particularly true of the sedimentary deposits where hematite forms by inorganic or biological precipitation in a body of water.

Minor clastic sedimentation can add clay minerals to the iron oxide. Episodic sedimentation can cause the deposit to have alternating bands of iron oxide and shale. Silica in the form of jasper, chert, or chalcedony can be added by chemical, clastic, or biological processes in small amounts or in significant episodes. These layered deposits of hematite and shale or hematite and silica have become known as the "banded iron formations".

Massive Hematite: A specimen of massive hematite about four inches across (ten centimeters) collected near Antwerp, New York.

Kidney Ore Hematite: Some hematite precipitates in cavities and has the opportunity to form an unrestricted habit. A habit known as "kidney ore" often develops in cavities and is named for its similar visual appearance to an internal organ. This type of chemically precipitated hematite is often relatively uncontaminated with sedimentary clay or host rock inclusions and has a higher purity. The high purity makes this the hematite of choice for making pigments. This specimen is about four inches across (ten centimeters) and was collected near Cumberland, England.

Geologic Occurrence

Hematite is found as a primary mineral and as an alteration product in igneous, metamorphic, and sedimentary rocks. It can crystallize during the differentiation of a magma or precipitate from hydrothermal fluids moving through a rock mass. It can also form during contact metamorphism when hot magmas react with adjacent rocks.

The most important hematite deposits formed in sedimentary environments. About 2.4 billion years ago, Earth's oceans were rich in dissolved iron, but very little free oxygen was present in the water. Then a group of cyanobacteria became capable of photosynthesis. The bacteria used sunlight as an energy source to convert carbon dioxide and water into carbohydrates, oxygen, and water. This reaction released the first free oxygen into the ocean environment. The new oxygen immediately combined with the iron to form hematite, which sank to the bottom of the seafloor and became the rock units that we know today as the banded iron formations.

Soon, photosynthesis was occurring in many parts of Earth's oceans, and extensive hematite deposits were accumulating on the seafloor. This deposition continued for hundreds of millions of years - from about 2.4 to 1.8 million years ago. This allowed the formation of iron deposits hundreds to several thousand feet thick that are laterally persistent over hundreds to thousands of square miles. They comprise some of the largest rock formations in Earth's rock record.

Many of the sedimentary iron deposits contain both hematite and magnetite as well as other iron minerals. These are often in intimate association, and the ore is mined, crushed, and processed to recover both minerals. Historically, much of the hematite was not recovered and was sent to tailings piles. More efficient processing today allows more hematite to be recovered from the ore. The tailings can also be reprocessed to recover additional iron and reduce tailings volume.

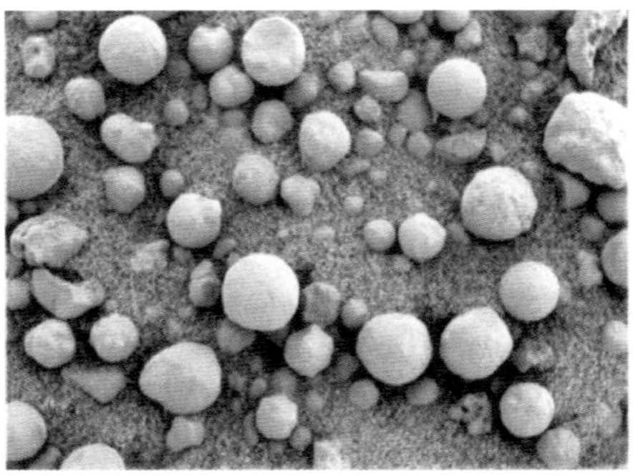

Martian "Blueberries": NASA's Mars Exploration Rover Opportunity discovered that soil near its landing site contained millions of tiny spheres that researchers nicknamed "blueberries." Upon analysis, they were determined to be composed of iron oxide, mostly in the form of hematite. The iron content of Martian rocks and soil contribute to its red appearance from Earth and helped it earn the name "The Red Planet."

Hematite on Mars

NASA has discovered that hematite is one of the most abundant minerals in the rocks and soils on the surface of Mars. An abundance of hematite in Martian rocks and surface materials gives the landscape a reddish brown color and is why the planet appears red in the night sky. It is the origin of Mars' "Red Planet" nickname.

Taconite Pellets: These taconite pellets consist of finely crushed taconite rock that has been processed to improve the iron content and mixed with a small amount of clay to improve pelletization. This is one of the standard ways of shipping iron ore from a mine to a steel mill. The round particles are about 1/2 inch in diameter (1 1/4 centimeter) and are very easy to handle during shipping and at the mill.

Uses of Hematite (Iron Ore)

Hematite is the world's most important ore of iron. Although magnetite contains a higher percentage of iron and is easier to process, hematite is the leading ore because it is more abundant and present in deposits in many parts of the world.

Hematite is mined in some of the largest mines in the world. These mines require investments of billions of dollars, and some will remove over 100 million tons of ore per year. These open-pit mines can be hundreds to thousands of feet deep and several miles across by the time they have been worked to completion.

China, Australia, Brazil, India, Russia, Ukraine, South Africa, and the United States are the world's leading producers of iron ore (includes hematite, magnetite, and other ores). Iron ore production in the United States occurs in Michigan and Minnesota.

Hematite Pigment: Hematite was one of the first pigment minerals used by people. At least 40,000

years ago, people obtained hematite, crushed it into a fine powder, and used it to make paints. Shown above are commercial hematite pigments that are available today. From top left, going clockwise, they are: Blue Ridge Hematite, Blue Ridge Violet Hematite, Venetian Red, and Pozzuoli Red. Since the Renaissance, pigments have often been named after the locations where they were produced. The color variations are a result of the type of hematite used and the impurities, such as clay and other iron oxides, that are commingled with it.

Hematite gems: Hematite and taconite are often made into tumbled stones or cut into cabochons and beads. These are popular as inexpensive jewelry items. Tumble-polished hematite is also popular as a "healing stone." Some people believe that carrying it will help relieve certain medical problems. This use has no scientific merit and can actually be harmful because it diverts people who need medical attention from seeing a doctor.

Uses of Hematite (Pigment)

The name hematite is from the Greek word "haimatitis" which means "blood-red." That name stems from the color of hematite when it has been crushed to a fine powder. Primitive people discovered that hematite could be crushed and mixed with a liquid for use as a paint or cosmetic. Cave paintings, known as "pictographs," dating back to 40,000 years ago were created with hematite pigments.

Hematite continues to be one of the most important pigment minerals. It has been mined at many locations around the world and has been traded extensively as a red pigment. During the Renaissance when many painters began using oils and canvas, hematite was one of the most important pigments. Hematite color was opaque and permanent. It could be mixed with a white pigment to produce a variety of pink colors that were used to paint flesh.

Uses of Hematite (Gem Material)

Hematite is a minor gem material used to produce cabochons, beads, small sculptures, tumbled stones, and other items. The material used to manufacture these products is a silver-colored hematite with a solid, uniform texture. The bright silver color of hematite and its "weighty feel" make it a very popular tumbled stone.

Hematite Novelties: Products called "magnetic hematite" and "iridescent hematite" are often offered for sale in gift, tourist, novelty, and science shops and their websites. Most of the time these materials are not hematite but are man-made materials that do not even have the same chemical composition as hematite. Buy them if you like them but not because you think that you are getting a unique mineral specimen.

Uses of Hematite (Healing Stone)

Some people believe that carrying pieces of tumble-polished hematite, known as "healing stones," will bring relief from certain medical problems. There is no scientific proof that this use of hematite has any positive effect beyond being a placebo. Using hematite as a "healing stone" or a "healing crystal" can actually be harmful because it diverts people from seeing a doctor who can provide proper care. Then when the person with the problem finally decides to see a doctor, their situation is more severe.

Iron Furnace: In the 1700s and 1800s, small mines in the eastern United States produced hematite which served as the primary iron ore of the region. The ore was processed by heating it by burning charcoal in simple stone furnaces. The iron ore deposits were small and difficult to exploit. When the large iron ore deposits of the Great Lakes region were discovered, iron ore was no longer mined in the eastern United States. Shown is the Vesuvius Iron Furnace of southern Ohio.

Other Uses of Hematite

Hematite is used for a number of other purposes. It is a very dense and inexpensive material that is effective at stopping x-rays. For that reason it is used for radiation shielding around medical and scientific equipment. The low cost and high density of hematite and other iron ores also makes them useful as ballast for ships.

Hematite can also be ground to a fine powder that when mixed with water will make a liquid with a very high specific gravity. These liquids are used in the "float-sink" processing of coal and other mineral material. The crushed coal, which has a very low specific gravity, is placed on the heavy liquid and the light clean coal floats, while high-specific-gravity impurities such as pyrite sink.

Finally, hematite is the material used to make polishing compounds known as "red rouge" and "jeweler's rouge." Red rouge is a hematite powder used to polish brass and other soft metals. It can be added to crushed corn cob media or crushed walnut shell media for tumble-polishing brass shell casings. Jeweler's rouge is a paste used on a soft cloth to polish gold and silver jewelry.

Kamacite

Kamacite is a nickel-rich variety of iron and is the principle constituent of a iron that is typical meteorite. Iron found on Earth is certainly caused by within the type of meteorites that have affected our planet's surface. There are very few resources of "native" iron. A couple of rare terrestrial gabbros (intrusive igneous rocks) and sulfide deposits do contain iron-nickel that is elemental. These are truly the only iron-nickel that is really native.

All iron that is natural whether it is indigenous or meteoritic, is actually an alloy of iron and nickel. The 2 elements are combined in varying percentages from significantly less than 6% nickel up to 75% nickel, although iron is through much more typical than nickel. Kamacite contains about 92% iron and 7% nickel. Another variety that is nickel-rich of meteorite is Taenite which typically contains about 25 – 40% iron.

Terrestrial occurrences of native iron are available at Blaafjeld, near Ovifak, Disko Island, and Greenland; in Germany, from Bühl, near Weimar, Hesse; on the Putorana Plateau, Taimyr Peninsula, Russia. Iron meteorites are found worldwide.

Physical properties

Kamacite has many unique physical properties including Thomson structures and extremely high density.

Identification

Kamacite is opaque, and its surface generally displays varying shades of gray streaking, or "quilting" patterns. Kamacite has a metallic luster. Kamacite can vary in hardness based on the extent of shock it has undergone, but commonly ranks a four on the mohs hardness scale. Shock increases kamacite hardness, but this is not 100% reliable in determining shock histories as there is a myriad of other reasons the hardness of kamacite could increase.

Kamacite has a measured density of 7.9 g/cm^3. It has a massive crystal habit but normally individual crystals are indistinguishable in natural occurrences. There are no planes of cleavage present in kamacite which gives it a hackly fracture. Kamacite is magnetic, and isometric which makes it behave optically isometrically.

Kamacite occurs with taenite and a mixed area of kamacite and taenite referred to as plessite.

Taenite contains more nickel (12 to 45 wt. % Ni) than kamacite (which has 5 to 12 wt. % Ni). The increase in nickel content causes taenite to have a face-centered unit cell, whereas kamacite's higher iron content causes its unit cell to be body centered. This difference is caused by nickel and iron having a similar size but different interatomic magnetic and quantum interactions.

Tetragonal Phase

There is evidence of a tetragonal phase, observed in X-ray powder tests and later under a microscope. When tested two meteorites gave d-values that could "be indexed on the basis of a tetragonal unit cell, but not on the basis of a cubic or hexagonal unit cell". It has been speculated to be e-iron, a hexagonal polymorph of iron.

Thomson Structures

Thomson structures, usually referred to as Widmanstätten patterns are textures often seen in meteorites that contain kamacite. These are bands which are usually alternating between kamacite and taenite. G. Thomson stumbled upon these structures in 1804 after cleaning a specimen with nitric acid he noticed geometric patterns. He published his observations in a French journal but due to the Napoleonic wars the English scientists, who were doing much of the meteorite research of the time, never saw his work. It was not until four years after in 1808 the same patterns were discovered by Count Alois von Beckh Widmanstätten who was heating iron meteorites when he noticed geometric patterns caused by the differing oxidation rates of kamacite and taenite. Widmanstätten told many of his colleagues about these patterns in correspondence leading to them being referred to as Widmanstätten patterns in most literature.

Thomson structures or Widmanstätten patterns are created as the meteorite cools; at high temperatures both iron and nickel have face-centered lattices. When the meteorite is formed it starts out as entirely molten taenite (greater than 1500 °C) and as it cools past 723 °C the primary metastable phase of the alloy changes into taenite and kamacite begins to precipitate out. It is in this window where the meteorite is cooling below 723 °C where the Thomson structures form and they can be greatly affected by the temperature, pressure, and composition of the meteorite.

Optical Properties

Kamacite is opaque and can be observed only in reflected light microscopy. It is isometric and therefore behaves isotropically.

Magnetism

As the meteorite cools below 750° C iron becomes magnetic as it moves into the kamacite phase. During this cooling the meteorite takes on non-conventional thermoremanent magnetization. Thermoremanent magnetization on Earth gives iron minerals formed in the Earth's crust, a higher magnetization than if they were formed in the same field at room temperature. This is a non-conventional thermoremanent magnetization because it appears to be due to a chemical remanent process which is induced as taenite is cooled to kamacite. What makes this especially interesting is this has been shown to account for all of the ordinary chondrites magnetic field which has been shown to be as strong as 0.4 Os.

Crystallography

Kamacite is an isometric mineral with a body centered unit cell. Kamacite is usually not found in large crystals; however the anomalously largest kamacite crystal found and documented measured 92×54×23 centimeters. Even with large crystals being so rare crystallography is extremely important to understand plays an important role in the formation of Thomson structures.

Symmetry

Kamacite forms isometric, hexoctahedral crystals this causes the crystals to have many symmetry elements. Kamacite falls under the 4/m32/m class in the Hermann–Mauguin notation meaning it has three fourfold axes, four threefold axes, and six twofold axes and nine mirror planes. Kamacite has a space group of F m3m.

Unit Cell

Kamacite is made up of a repeating unit of α-(Fe, Ni); Fe0+0.9Ni0.1 which makes up cell dimensions of a = 8.603, Z = 54; V = 636.72. The interatomic magnetic and quantum interactions of the iron atoms interacting with each other causes kamacite to have a body centered lattice.

Chemistry

Formula and Dominant Elements

Kamacite is made up of a repeating unit of α-(Fe, Ni); Fe0+0.9Ni0.1. Besides trace elements, it is normally considered to be made up of 90% iron and 10% nickel but can have a ratio of 95% iron and 5% nickel. This makes iron the dominant element in any sample of kamacite. It is grouped with the native elements in both Dana and Nickel-Strunz classification systems.

Conditions of Formation

Kamacite starts to form around 723 °C, where iron splits from being face centered to body centered while nickel remains face centered. To accommodate this areas start to form of higher iron concentration displacing nickel to the areas around it which creates taenite which is the nickel end member.

Trace Elements

There has been a great deal of research into kamacite's trace elements. The most notable trace elements in kamacite are gallium, germanium, cobalt, copper, and chromium. Cobalt is the most notable of these where the nickel content varies from 5.26% to 6.81% and the cobalt content can be from 0.25% to 0.77%. All of these trace elements are metallic and their appearance near the kamacite taenite border can give important clues to the environment the meteorite was formed in. Mass spectroscopy has revealed kamacite to contain considerable amounts of platinum to be an average of 16.31 $(\mu g / g)$, iridium to be an average of 5.40 (µg/g), osmium to be an average of 3.89 $(\mu g / g)$, tungsten to be an average of 1.97 $(\mu g / g)$, gold to be an average of 0.75 $(\mu g / g)$, rhenium to be an average of 0.22 (µg/g). The considerable amounts of cobalt and platinum are the most notable.

Important Minor Elements, Substitutions and Solid Solutions

Kamacite sulfurization has been done experimentally in laboratory conditions. Sulfurization resulted in three distinct phases: a mono-sulfide solid solution (Fe,Ni,Co)1-xS, a pentlandite phase $(Fe,Ni,Co)_9$-xS_8, as well as a P-rich phase. This was done in a lab to construct conditions concurrent with that of the solar nebula. With this information it would be possible to extract information about the thermodynamic, kinetic, and physical conditions of the early solar system. This still remains speculatory as many of the sulfides in meteorites are unstable and have been destroyed. Kamacite also alters to tochilinite (Fe_2+5-6(Mg, Fe_2+$)_5S_6(OH)_{10}$). This is useful for giving clues as to how much the meteorite as a whole has been altered. Kamacite to tochilinite alteration can be seen in petrologic microscopes, scanning electron microscope, and electron microprobe analysis. This can be used to allow researchers to easily index the amount of alteration that has taken place in the sample. This index can be later referenced when analyzing other areas of the meteorite where alteration is not as clear.

Relationship with Taenite

Taenite is the nickel rich end member of the kamacite–taenite solid solution. Taenite is naturally occurring on Earth whereas kamacite is only found on Earth when it comes from space. Kamacite forms taenite as it forms and expels nickel to the surrounding area, this area forms taenite. Due to the face centered nature of the kamacite lattice and the body centered nature of the nickel lattice the two make intricate angles when they come in contact with each other. These angles reveal themselves macroscopically in the Thomson structure. Also due to this relationship we get the terms ataxite, hexahedrites and octahedrite. Ataxite refers to meteorites that do not show a grossly hexahedral or octahedral structure. Meteorites composed of 6 wt% or less nickel are often referred to as hexahedrites due to the crystal structure of kamacite being isometric and causing the meteorite to be cubic. Likewise if the meteorite is dominated by the face centered taenite it is called an octahedrite as kamacite will exsolve from the octahedral crystal boundaries of taenite making the meteorite appear octahedral. Both hexahedrites and octahedrite only appear when the meteorite breaks along crystal planes or when prepared to excentuate the Thomson structures therefore many are mistakenly called ataxites ar first.

Chemical Explanation of Heat

Trace elements have been analyzed in the formation of kamacite at different temperatures but the trace elements in taenite seem better suited to give clues of the formation temperature of the meteorite. As the meteorite cools and taenite and kamacite are sorting out of each other some of the trace elements will prefer to be located in taenite or kamacite. Analyzing the taenite kamacite boundary can give clues to how quickly cooling occurred as well as a myriad of other conditions during formation by the final location of the trace elements.

Stability Range

Kamacite is only stable at temperatures below 723 °C or 600°C as that is where iron becomes cool enough to arrange in a body centered arrangement. Kamacite is also only stable at low pressures as can be assumed as it only forms in space.

Effect of Shock

Metallographic and X-ray diffraction can be used on kamacite to determine the shock history of a meteorite. Using hardness to determine shock histories has been experimented with but was found to be too unreliable. Vickers hardness test was applied to a number of kamacite samples and shocked meteorites were found to have values of 160–170 kg/mm and non-shocked meteorites can have values as high as 244 kg/mm. Shock causes a unique iron transformation structure that is able to be measured using metallographic and X-ray diffraction techniques. After using metallographic and X-ray diffraction techniques to determine shock history it was found that 49% of meteorites found on Earth contain evidence of shock.

Geologic Occurrences

Kamacite meteorites has been found on every continent on Earth and it has also been found on Mars.

Meteorites

Kamacite is primarily associated with meteorites because it needs high temperatures, low pressures and few other more reactive elements like oxygen. Chondrite meteorites can be split into groups based on the chondrules present. There are three major types: enstatite chondrites, carbonaceous chondrites and ordinary chondrites. Ordinary chondrites are the most abundant type of meteorite found on Earth making up 85% of all meteorites recorded. Ordinary chondrites are thought to have all originated from three different sources thus they come in three types LL, L, and H; LL stands for Low iron, Low metal, L stands for Low iron abundance, and H is High iron content. All ordinary chondrites contain kamacite in decreasing abundance as you move from H to LL chondrites. Kamacite is also found in many of the less common meteorites mesosiderites and E chondrites. E chondrites are chondrites which are made primarily of enstatite and only account for 2% of meteorites that fall onto the Earth. E chondrites have an entirely different source rock than that of the ordinary chondrites. In analysis of kamacite in E chondrites it was found that they contain generally less nickel then average.

Abundance

Since kamacite is only formed in space and is only found on Earth in meteorites, it has very low abundance on Earth. Its abundance outside our solar system is difficult to determine. Iron, the main component of kamacite, is the sixth most abundant element in the universe and the most abundant of those elements generally considered metallic.

Associated Minerals

Taenite, and tochilinite are minerals that are commonly associated with kamacite.

Specific Examples

Meteor Crater Arizona

Kamacite has been found and studied in Meteor Crater, Arizona. Meteor Crater was the first confirmed meteor impact site on the planet, and was not universally recognized as such till the 1950s. In the 1960s United States Geological Survey discovered kamacite in specimens gathered from around the site tying the mineral to meteorites.

Planets

Kamacite primarily forms on meteorites but has been found on extraterrestrial bodies such as Mars. This was discovered by The Mars Exploration Rover (MER) Opportunity. The kamacite did not originate on Mars but was put there by a meteorite. This was particularly of interest because the meteorite fell under the lesser known class of mesosiderites. Mesosiderites are very rare on Earth and its occurrence on Mars gives clues to the origin of its larger source rock.

Uses

The primary research use of kamacite is to shed light on a meteorite's history. Whether it is looking at the shock history in the iron structures or the conditions during the formation of the meteorite using the kamacite-taenite boundary understanding kamacite is key to understanding our universe.

Museums, University and Photo Specimen Preparation

Due to the rareness and the generally dull appearance of kamacite it is not popular among private collectors. However many museums and universities have samples of kamacite in their collection. Normally kamacite samples are prepared using polish and acid to show off the Thomson structures. Preparing specimens involves washing them in a solvent, such as Thomson did with nitric acid to bring out the Thomson structures. Then they are heavily polished so they look shiny. Generally the kamacite can be told apart from taenite easily as after this process the kamacite looks slightly darker than the taenite.

Looking to the Future

Kamacite and taenite both have the potential to be economically valuable. An option that would make asteroid mining more profitable would be to gather the trace elements. One difficulty would be refining elements such as platinum and gold. Platinum is worth around 12,000 US$/kg and (kamacite contains 16.11 μg/g platinum) and gold is worth around 12,000 US$/kg (kamacite contains 0.52 μg/g gold); however the likeliness of a profitable return is fairly slim. Asteroid mining for space uses could be more practical, as transporting materials from Earth is costly. Similar to current plans of reusing the modules of the International Space Station in other missions, an iron meteorite could be used to build space craft in space. NASA has put forward preliminary plans to build a space ship in space.

Ilmenite

Ilmenite is a common accessory mineral in igneous rocks, sediments, and sedimentary rocks in many parts of the world. Apollo astronauts found abundant ilmenite in lunar rocks and the lunar regolith. Ilmenite is a black iron-titanium oxide with a chemical composition of $FeTiO_3$.

Ilmenite: A specimen of massive ilmenite from Saint-Urbain, Quebec, Canada. Massive ilmenite can be formed as a vein-filling material or during magmatic segregation. This specimen is approximately 4 inches (10 centimeters) across.

Ilmenite is the primary ore of titanium, a metal needed to make a variety of high-performance alloys. Most of the ilmenite mined worldwide is used to manufacture titanium dioxide, TiO_2, an important pigment, whiting, and polishing abrasive.

Heavy Mineral Sand: Shallow digging at Folly Beach, South Carolina, exposes thin layers of heavy-mineral sands. Most of the ilmenite mined today is from sands with a heavy mineral concentration.

Mining Heavy Minerals: Excavators remove heavy mineral sands at the Concord Mine in south-central Virginia. Weakly consolidated sands containing about 4% heavy minerals are excavated and processed to remove ilmenite, leucoxene, rutile, and zircon. The sands were weathered and eroded from an anorthocite exposure a short distance away.

Geologic Occurrence

Most ilmenite forms during the slow cooling of magma chambers and is concentrated through the process of magmatic segregation. A large underground magma chamber can take centuries to cool. As it cools, crystals of ilmenite will begin forming at a specific temperature. These crystals are heavier than the surrounding melt and sink to the bottom of the magma chamber.

This causes ilmenite and similar-temperature minerals, such as magnetite, to accumulate in a layer at the bottom of the magma chamber. These ilmenite-bearing rocks are often gabbro, norite, or anorthosite. Ilmenite also crystallizes in veins and cavities and sometimes occurs as well-formed crystals in pegmatites.

Ilmenite has a high resistance to weathering. When rocks containing ilmenite weather, grains of ilmenite disperse with the sediment. The high specific gravity of these grains causes them to segregate during stream transport and accumulate as "heavy mineral sands." These sands are black in color and easily recognized by geologists. "Black sand prospecting" has long been a method of finding heavy mineral placer deposits. Most commercially produced ilmenite is recovered by excavating or dredging these sands, which are then processed to remove the heavy mineral grains such as ilmenite, leucoxene, rutile, and zircon.

Ilmenite: A specimen of massive ilmenite from Normanville, South Australia. Specimen is approximately 3 inches (7.6 centimeters) across.

Chemical Composition of Ilmenite

Ilmenite's ideal chemical composition is $FeTiO_3$. However, it often departs from that composition by containing variable amounts of magnesium or manganese. These elements substitute for iron in complete solid solution. A solid solution series exists between ilmenite ($FeTiO_3$) and geikielite ($MgTiO_3$). In this series, variable amounts of magnesium substitutes for iron in the mineral's crystal structure. A second solid solution series exists between ilmenite and pyrophanite ($MnTiO_3$), with manganese substituting for iron. At high temperatures, a third solid solution series exists between ilmenite and hematite (Fe_2O_3).

Ilmenite: A specimen of massive ilmenite from Kragero, Norway. Specimen is approximately 4 inches (10 centimeters) across.

Black Sand Ilmenite: Ilmenite sand from Melbourne, Florida. Specimens are sand-size grains.

Physical Properties of Ilmenite

Ilmenite is a black mineral with a submetallic to metallic luster. With just a glance it can easily be confused with hematite and magnetite. The differentiation is easy. Hematite has a red streak, while ilmenite has a black streak. Magnetite is strongly magnetic, while ilmenite is not magnetic. Occasionally ilmenite is weakly magnetic, possibly from small amounts of included magnetite.

Physical Properties of Ilmenite	
Chemical Classification	Oxide
Color	Black
Streak	Black
Luster	Metallic, submetallic

Diaphaneity	Opaque
Cleavage	None
Mohs Hardness	5.5 to 6
Specific Gravity	4.7 to 4.8
Diagnostic Properties	Streak; sometimes weakly magnetic.
Chemical Composition	Iron titanium oxide - $FeTiO_3$. Sometimes has significant amounts of magnesium and manganese in solid solution with the iron to yield a composition of $(Fe,Mg,Mn)TiO_3$
Crystal System	Hexagonal
Uses	The primary ore of titanium. A minor source of iron. Used to make titanium dioxide.

Ilmenite is usually more durable than the other minerals in the igneous rocks in which it is abundant. For that reason, the weathering debris produced during the weathering of these rocks is especially rich in ilmenite. Its relatively high specific gravity causes it to become concentrated in placer deposits like gold, gems, and other heavy minerals.

Pigments and Polishing Compounds: Titanium dioxide powder is carefully processed to remove impurities and classified by particle size. It is then sold for use as whiting, pigments, and polishing compounds. The image is a rock tumbler barrel just opened with a thick white froth of metal oxide polish.

Lunar Ilmenite Basalt: Apollo astronauts found ilmenite-rich basalts at multiple locations on the Moon. The reference block at lower right is one cubic centimeter.

Uses of Ilmenite

Ilmenite is the primary ore of titanium metal. Small amounts of titanium combined with certain metals will produce durable, high-strength, lightweight alloys. These alloys are used to manufacture a wide variety high-performance parts and tools. Examples include: aircraft parts, artificial joints for humans, and sporting equipment such as bicycle frames. About 5% of the ilmenite mined is used to produce titanium metal. Some ilmenite is also used to produce synthetic rutile, a form of titanium dioxide used to produce white, highly reflective pigments.

Most of the remaining ilmenite is used to make titanium dioxide, an inert, white, highly reflective material. The most important use of titanium dioxide is as a whiting. Whitings are white, highly reflective materials that are ground to a powder and used as pigments. These pigments produce a white color and brightness in paint, paper, adhesives, plastics, toothpaste, and even food.

Titanium dioxide is also used to make powders with a tightly controlled particle size range. These powders are used as inexpensive polishing abrasives in a variety of lapidary work that includes rock tumbling, lapping, cabbing, sphere making, and faceting. Titanium oxide abrasives are used in many other industries.

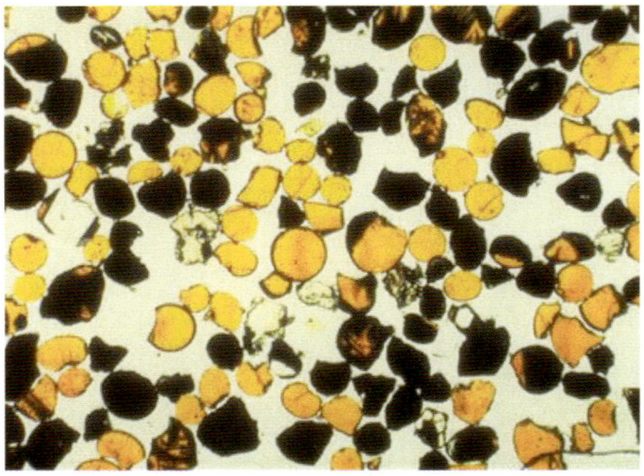

Lunar Ilmenite Regolith: Apollo astronauts found deposits of lunar regolith composed mostly of silt- to sand-size ilmenite (black) and mafic volcanic glass (orange).

Ilmenite on the Moon

Apollo astronauts found ilmenite-rich basalts at multiple locations on the Moon. Most of these basalts were extremely old, forming at least 3 billion years ago. These rocks often contained over 10% titanium dioxide (TiO_2). Minerals present in these rocks were mostly feldspars and pyroxenes, with ilmenite next in abundance.

Some samples of lunar regolith contained significant amounts of ilmenite. It occurred in particles ranging from fine silt to coarse sand. The ilmenite was thought to have been liberated from lunar basalts during impact events.

Samples of lunar regolith collected at Shorty Crater contained a mix of volcanic glass spheres and ilmenite grains. The deposit was stratified with a bottom layer composed mostly of ilmenite and other black opaque materials. This graded upwards to an upper layer, known as "orange soil," that was composed mostly of spherical-shaped beads of orange volcanic glass with minor amounts of ilmenite. The grains were mostly less than 1/2 millimeter in size. This regolith was thought to have been produced by fountaining volcanic eruptions during the early lunar history.

References

- Lauretta, D. (1998). "Kamacite sulfurization in the solar nebula". Meteoritics & Planetary Science. 33: 4. Bibcode:1998M&PS...33..821L. doi:10.1111/j.1945-5100.1998.tb01689.x.

- Magnetite: newworldencyclopedia.org, Retrieved 22 May 2018
- Flemming, R. (2007). "Micro X-ray diffraction (μXRD): a versatile technique for characterization of Earth and planetary materials". Canadian Journal of Earth Sciences. 44(9): 1333–1346. Bibcode:2007CaJES..44.1333F. doi:10.1139/e07-020.
- Maghemite-powder, products: reade.com, Retrieved 12 April 2018
- Easton, A. J. (1986). "Studies of kamacite, perryite and schreibersite in E-chondrites and aubrites". Meteoritics. 21: 79–93. doi:10.1111/j.1945-5100.1986.tb01227.x.
- Kamacite: nationalgemlab.in, Retrieved 30 June 2018

Chapter 5

Gemology: A Study of Gemstone Minerals

Gemology is a branch of mineralogy, which is concerned with artificial and natural gemstones. This chapter has been carefully written to provide an easy understanding of gemology and the study of gemstone minerals. Many topics related to this field such as cardinal gem, asterism and chatoyancy have been extensively explored in this chapter.

Gemology is the combined art and science of studying, cutting, valuing, buying and selling precious stones. Some of the most precious jewels that gemologists deal in include diamonds, rubies, sapphires, and emeralds. Gemology is a type of mineralogy that specifically focuses on the study of precious gems, both natural and those composed of artificial materials. Some jewelers are gemologists. Appraisers who evaluate the value of gemstones should also be a gemologist or at least have studied extensively to master the most important aspects of determining a gem's condition and value.

With advances in gemstone synthesis, gemology has become a much more important field of study. Often, industry professionals are the most highly trained and educated gemologists. These professionals have become certified gemologists. An accredited gemologist is an expert in the field who has completed an intense training and gone through advanced education in various gemstone areas. Often, they will specialize in a specific area of the business or a particular type of stone. Many gemologists specialize in the study and evaluation of diamonds.

Gemologists value stones based on factors, including cut, color, quality, and clarity. Many people are familiar with a similar group of criteria for gemstone evaluation used in assessing diamonds. For diamonds, the factors are generally known as the 4 Cs of color, clarity, cut and carat.

Gemology and Gemstone Investing

When returns in the stock market decline, aggressive investors often seek out alternative investments which hold more promise of increasing returns on invested capital. Precious gems, in particular, those that are rare or of exceptional quality, have a high likelihood of at least retaining their value, and are likely to increase in value. However, they are not an asset which may be as easily liquidated as other investments if there is an urgent need for cash. This problem is especially real for rare, precious stones which would appeal to an elite level of buyer.

Gemstone investing can seem exciting to people who want to make quick returns, but it is highly speculative and should only be undertaken by experienced professionals. The term investment-grade is often tossed around by those who want to sell gems or try to convince other people to invest in them. However, this is controversial and frowned upon by industry leaders, as there are no formal standards for what constitutes an investment-grade gemstone.

Gemstone

A gemstone is the name given to a piece of mineral (or other rock or organic material) that after it has been cut and polished has been made into a piece of jewellery or another accessory. Although gemstones are commonly made from minerals, material such as jet or amber or rocks like lapis lazuli can also be utilised in the creation of gemstones. The majority of gemstones will be hard to the touch but depending on the piece of jewellery that is being created, some soft minerals are also utilised. Take a look at Moh's Scale of Hardness to find out more about this.

Some other names for a gemstone include:

- Gem/Fine Gem
- Precious stone/Semi-precious stone
- Jewel

Formation of Gemstones

Even though the vast majority of gemstones are created below the surface of the earth, there are a number of different ways that the individual stones can be created. A great number of these precious stones are created when the minerals react with water that is found just below the surface of the earth and the minerals then dissolve. This process allows the minerals to form with gemstones like opals, agates and amethysts forming when the solution cools or evaporates.

Gems like malachite, azurite or turquoise are formed when water blends with rocks that are rich in copper. There is a strong history attached to turquoise gems with the ancient Egyptians regarding them to be an extremely precious gem. This means that turquoise gems have been around for 4,000 years or even longer.

Value of Gemstones

Much the same as you would expect from any other major industry, the price of gemstones is largely influenced by supply and demand.

One of the most important things for people when considering gemstones is the value and price for them but like in many industries, the laws of supply and demand are very much in place. This means that rare gems will carry a higher price tag.

For instance, a garnet is extremely stylish and attractive but it is not that rare, which means that its price tag is not too high. If you compared a garnet to a ruby of a similar size and condition, the ruby would always hold a higher price, simply because of the fact that it is a rare stone.

It is often difficult to place a true value on any particular gemstone because there is no uniform classification or measurement system in place. This is where gemstones differ from diamonds, which have been subject to a uniform measurement system since the 1950s. Traditionally, gemstones have been examined and evaluated by eye. When examining a gemstone, the four most important factors can be summed up in the four C's, which are; colour, clarity, cut and carat.

Even though diamonds are one of the most prized possessions on Earth, there are actually a number of different gemstones that are rarer to find than diamonds. Some of the rarest gemstones you can find include:

- Painite
- Alexandrite
- Tanzanite
- Benitoite
- Poudretteite
- Grandiderite
- Musgravite
- Jeremejevite
- Red Beryl

Cutting and Polishing

A few gemstones are used as gems in the crystal or other form in which they are found. Most however, are cut and polished for usage as jewelry. The picture to the left is of a rural, commercial cutting operation in Thailand. This small factory cuts thousands of carats of sapphire annually. The two main classifications are stones cut as smooth, dome shaped stones called cabochons, and stones which are cut with a faceting machine by polishing small flat windows called facets at regular intervals at exact angles.

Raw gemstones

A diamond cutter in Amsterdam in the Netherlands in 2012

Stones which are opaque or semi-opaque such as opal, turquoise, variscite, etc. are commonly cut as cabochons. These gems are designed to show the stone's color or surface properties as in opal and star sapphires. Grinding wheels and polishing agents are used to grind, shape and polish the smooth dome shape of the stones.

Gems which are transparent are normally faceted, a method which shows the optical properties of the stone's interior to its best advantage by maximizing reflected light which is perceived by the viewer as sparkle. There are many commonly used shapes for faceted stones. The facets must be cut at the proper angles, which varies depending on the optical properties of the gem. If the angles are too steep or too shallow, the light will pass through and not be reflected back toward the viewer. The faceting machine is used to hold the stone onto a flat lap for cutting and polishing the flat facets. Rarely, some cutters use special curved laps to cut and polish curved facets.

Colors

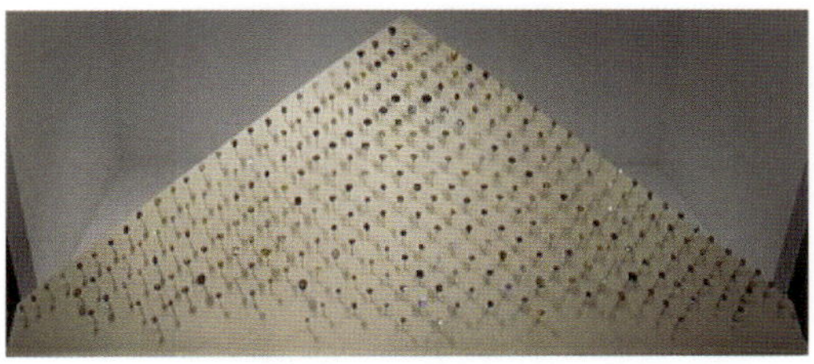

Nearly 300 variations of diamond color exhibited at the Aurora display at the Natural History Museum in London.

The color of any material is due to the nature of light itself. Daylight, often called white light, is actually all of the colors of the spectrum combined. When light strikes a material, most of the light is absorbed while a smaller amount of a particular frequency or wavelength is reflected. The part that is reflected reaches the eye as the perceived color. A ruby appears red because it absorbs all the other colors of white light (green and blue), while reflecting the red.

A material which is mostly the same can exhibit different colors. For example, ruby and sapphire have the same primary chemical composition (both are corundum) but exhibit different colors because of impurities. Even the same named gemstone can occur in many different colors: sapphires show different shades of blue and pink and "fancy sapphires" exhibit a whole range of other colors from yellow to orange-pink, the latter called "padparadscha sapphire".

This difference in color is based on the atomic structure of the stone. Although the different stones formally have the same chemical composition and structure, they are not exactly the same. Every now and then an atom is replaced by a completely different atom, sometimes as few as one in a million atoms. These so-called impurities are sufficient to absorb certain colors and leave the other colors unaffected.

For example, beryl, which is colorless in its pure mineral form, becomes emerald with chromium impurities. If manganese is added instead of chromium, beryl becomes pink morganite. With iron, it becomes aquamarine.

Some gemstone treatments make use of the fact that these impurities can be "manipulated", thus changing the color of the gem.

Treatment

Gemstones are often treated to enhance the color or clarity of the stone. Depending on the type and extent of treatment, they can affect the value of the stone. Some treatments are used widely because the resulting gem is stable, while others are not accepted most commonly because the gem color is unstable and may revert to the original tone.

Heat

Heat can improve gemstone color or clarity. The heating process has been well known to gem miners and cutters for centuries, and in many stone types heating is a common practice. Most citrine is made by heating amethyst, and partial heating with a strong gradient results in "ametrine" – a stone partly amethyst and partly citrine. Aquamarine is often heated to remove yellow tones, or to change green colors into the more desirable blue, or enhance its existing blue color to a purer blue.

Nearly all tanzanite is heated at low temperatures to remove brown undertones and give a more desirable blue/purple color. A considerable portion of all sapphire and ruby is treated with a variety of heat treatments to improve both color and clarity.

When jewelry containing diamonds is heated (for repairs) the diamond should be protected with boric acid; otherwise the diamond (which is pure carbon) could be burned on the surface or even burned completely up. When jewelry containing sapphires or rubies is heated, those stones should not be coated with boracic acid (which can etch the surface) or any other substance. They do not have to be protected from burning, like a diamond (although the stones *do* need to be protected from heat stress fracture by immersing the part of the jewelry with stones in water when metal parts are heated).

Radiation

Virtually all blue topaz, both the lighter and the darker blue shades such as "London" blue, has been irradiated to change the color from white to blue. Most greened quartz (Oro Verde) is also irradiated to achieve the yellow-green color. Diamonds are irradiated to produce fancy-color diamonds (which occur naturally, though rarely in gem quality).

Waxing/oiling

Emeralds containing natural fissures are sometimes filled with wax or oil to disguise them. This wax or oil is also colored to make the emerald appear of better color as well as clarity. Turquoise is also commonly treated in a similar manner.

Fracture Filling

Fracture filling has been in use with different gemstones such as diamonds, emeralds and sapphires. In 2006 "glass filled rubies" received publicity. Rubies over 10 carats (2 g) with large fractures were filled with lead glass, thus dramatically improving the appearance (of larger rubies in particular). Such treatments are fairly easy to detect.

Synthetic and Artificial Gemstones

It is important to distinguish between synthetic gemstones, and imitation or simulated gems.

Synthetic gems are physically, optically and chemically identical to the natural stone, but are created in controlled conditions in a laboratory. Imitation or simulated stones are chemically different than the natural stone but may be optically similar to it; they can be glass, plastic, resins or other compounds.

Examples of simulated or imitation stones include cubic zirconia, composed of zirconium oxide and simulated moissanite, which are both diamond simulants. The imitations copy the look and color of the real stone but possess neither their chemical nor physical characteristics. Moissanite actually has a higher refractive index than diamond and when presented beside an equivalently sized and cut diamond will have more "fire" than the diamond.

Synthetic, cultured or lab-created gemstones are not imitations. For example, diamonds, rubies, sapphires and emeralds have been manufactured in labs to possess identical chemical and physical characteristics to the naturally occurring variety. Synthetic (lab created) corundum, including ruby and sapphire, is very common and costs much less than the natural stones. Smaller synthetic diamonds have been manufactured in large quantities as industrial abrasives, although larger gem-quality synthetic diamonds are becoming available in multiple carats.

Whether a gemstone is a natural stone or lab-created (synthetic), the physical characteristics are the same. Lab-created stones tend to have a more vivid color to them, as impurities are not present in a lab and do not modify the clarity or color of the stone, unless added intentionally for a specific purpose.

Cardinal Gem

The term cardinal gemstone was traditionally used to refer to gemstones that were valued above all others. The five gemstones referred to as the cardinal gems included diamond, sapphire, ruby, emerald, and amethyst. The important use of these gemstones in traditional religious and cultural ceremonies and other events aided in labeling them as the cardinal gems. However, with the discovery of newer gemstones and new deposits of the cardinal gems in recent times, the term cardinal gems has largely fallen out of use.

Diamond (Gemstone)

Diamond is a mineral composed of pure carbon. It is the hardest naturally occurring substance known; it is also the most popular gemstone. Because of their extreme hardness, diamonds have a number of important industrial applications.

Diamond (Gem)

country	mine production 2006 (carats)*	% of world mine production
*Estimate.		
**Detail does not add to total given because of rounding.		
Australia	25,000,000	29.4
Congo (Kinshasa)	24,000,000	28.2
Russia	15,000,000	17.6
South Africa	9,000,000	10.6
Botswana	8,000,000	9.4
China	1,000,000	1.2
United States	300,000	0.4
Other countries	3,000,000	3.5
World total	85,000,000	100**

The hardness, brilliance, and sparkle of diamonds make them unsurpassed as gems. In the symbolism of gemstones, the diamond represents steadfast love and is the birthstone for April. Diamond stones are weighed in carats (1 carat = 200 milligrams) and in points (1 point = 0.01 carat). In addition to gem-quality stones, several varieties of industrial diamonds occur, and syntheticdiamonds have been produced on a commercial scale since 1960.

Diamonds are found in three types of deposits: alluvial gravels, glacial tills, and kimberlite pipes. The kimberlite pipes (such as those at Kimberley, South Africa) form from intrusions of magma into the Earth's crust and deliver diamonds and other rocks and minerals from the mantle. The pipes themselves are often less than 100 million years old. However, the diamonds they carry were formed 1 to 3.3 billion years ago at depths of more than about 75 miles (120 km). Diamonds found in alluvial and glacial gravels must have been released by fluvial or glacial erosion of the kimberlite matrix and then redeposited in rivers or in glacial till.

Alluvial diamond mining at Kenema, Sierra Leone *Bavaria-Verlag*

Alluvial diamond mine Large alluvial diamond mine near the Bow River in Western Australia.

A diamond mine at Kimberley, S.Af.

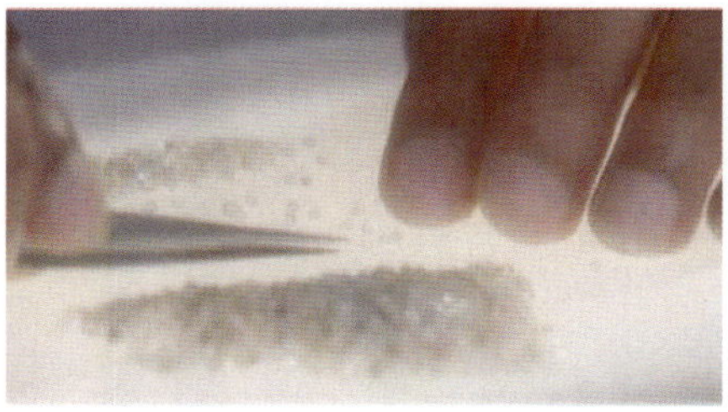

South Africa: diamond mining History of diamond mining in South Africa.

Diamonds vary from colourless to black, and they may be transparent, translucent, or opaque. Most diamonds used as gems are transparent and colourless or nearly so. Colourless or pale blue stones are most valued, but these are rare; most gem diamonds are tinged with yellow. A "fancy" diamond has a distinct body colour; red, blue, and green are rarest, and orange, violet, yellow, and yellowish green more common. Most industrial diamonds are gray or brown and are translucent or opaque, but better-quality industrial stones grade imperceptibly into poor quality gems. The colour of diamonds may be changed by exposure to intense radiation (as released in a nuclear reactor or by a particle accelerator) or by heat treatment.

A very high refractive power gives the diamond its extraordinary brilliance. A properly cut diamond will return a greater amount of light to the eye of the observer than will a gem of lesser refractive power and will thus appear more brilliant. The high dispersion gives diamonds their fire, which is caused by the separation of white light into the colours of the spectrum as it passes through the stone.

The scratch hardness of diamond is assigned the value of 10 on the Mohs scale of hardness; corundum, the mineral next to diamond in hardness, is rated as 9. Actually, diamond is very much harder than corundum; if the Mohs scale were linear, diamond's value would be about 42. The hardness of a diamond varies significantly in different directions, causing cutting and polishing of some faces to be easier than others. For detailed physical properties.

In the atomic structure of diamond, as determined by X-ray diffraction techniques, each carbon atomis linked to four equidistant neighbours throughout the crystal. This close-knit, dense, strongly bonded crystal structure yields diamond properties that differ greatly from those of graphite, native carbon's other form.

The most familiar usage of diamonds today is as gemstones used for adornment—a usage which dates back into antiquity. The dispersion of white light into spectral colors is the primary gemological characteristic of gem diamonds. In the twentieth century, gemologists have developed methods of grading diamonds and other gemstones based on the characteristics most important to their value as a gem. Four characteristics known informally as the "four Cs" are now commonly used as the basic descriptors of diamonds: carat, cut, color, and clarity. This system was developed by Gemological Institute of America in 1953 as internationally recognized standard to evaluate diamonds characteristics.

Most gem diamonds are traded on the wholesale market based on single values for each of the four Cs; for example knowing that a diamond is rated as 1.5 carats (300 mg), VS2 clarity, F color, excellent cut round brilliant, is enough to reasonably establish an expected price range. More detailed information from within each characteristic is used to determine actual market value for individual stones. Consumers who purchase individual diamonds are often advised to use the four Cs to pick the diamond that is "right" for them.

Other characteristics also influence the value and appearance of a gem diamond. These include physical characteristics such as the presence of fluorescence as well as the diamond's source and which gemological institute evaluated the diamond. *Cleanliness* also dramatically affects a diamond's beauty.

There are two major non-profit gemological associations which grade and provide reports, (informally referred to by the term *certificate* or *cert*, which is a misnomer for many grading reports) on diamonds; while carat weight and cut angles are mathematically defined, the clarity and color are judged by the trained human eye and are therefore open to slight variance in interpretation. These associations are listed below:

- Gemological Institute of America (GIA) was the first laboratory in America to issue modern diamond reports, and is held in high regard amongst gemologists for its consistent, conservative grading.
- Diamond High Council (HRD) Official certification laboratory of the Belgian diamond industry, located in Antwerp.

Within the last two decades, a number of for-profit gemological grading laboratories have also been established, many of them also based in Antwerp or New York. These entities serve to provide similar services as the non-profit associations above, but in a less expensive and more timely fashion. They produce certificates that are similar to those of the GIA.

Carat

The *carat weight* measures the mass of a diamond. One carat is defined as 200 milligrams (about 0.007 ounces avoirdupois). The *point* unit—equal to one one-hundredth of a carat (0.01 carat, or

2 mg)—is commonly used for diamonds of less than one carat. All else being equal, the price per carat increases with carat weight, since larger diamonds are both rarer and more desirable for use as gemstones.

The price per carat does not increase linearly with increasing size. Instead, there are sharp jumps around milestone carat weights, as demand is much higher for diamonds weighing just more than a milestone than for those weighing just less. As an example, a 0.99-carat diamond may have a significantly lower price per carat than a comparable 1.01-carat diamond, because of differences in demand.

A weekly diamond price list, the Rapaport Diamond Report is published by Martin Rapaport, CEO of Rapaport Group of New York, for different diamond cuts, clarity and weights. It is currently considered the de facto retail price baseline. Jewelers often trade diamonds at negotiated discounts off the Rapaport price (e.g., "R −3%").

In the wholesale trade of gem diamonds, carat is often used in denominating lots of diamonds for sale. For example, a buyer may place an order for 100 carats (20 g) of 0.5-carat (100 mg), D–F, VS2-SI1, excellent cut diamonds, indicating a wish to purchase 200 diamonds (100 carats (20 g) total mass) of those approximate characteristics. Because of this, diamond prices (particularly among wholesalers and other industry professionals) are often quoted per carat, rather than per stone.

Total carat weight (t.c.w.) is a phrase used to describe the total mass of diamonds or other gemstone in a piece of jewelry, when more than one gemstone is used. Diamond solitaire earrings, for example, are usually quoted in t.c.w. when placed for sale, indicating the mass of the diamonds in both earrings and not each individual diamond. T.c.w. is also widely used for diamond necklaces, bracelets and other similar jewelry pieces.

Clarity

Clarity is a measure of internal defects of a diamond called *inclusions*. Inclusions may be crystals of a foreign material or another diamond crystal, or structural imperfections such as tiny cracks that can appear whitish or cloudy. The number, size, color, relative location, orientation, and visibility of inclusions can all affect the relative clarity of a diamond. The Gemological Institute of America (GIA) and other organizations have developed systems to grade clarity, which are based on those inclusions which are visible to a trained professional when a diamond is viewed under 10× magnification.

Diamonds become increasingly rare when considering higher clarity gradings. Only about 20% of all diamonds mined have a clarity rating high enough for the diamond to be considered appropriate for use as a gemstone; the other 80% are relegated to industrial use. Of that top 20%, a significant portion contains one or more visible inclusions. Those that do not have a visible inclusion are known as "eye-clean" and are preferred by most buyers, although visible inclusions can sometimes be hidden under the setting in a piece of jewelry.

Most inclusions present in gem-quality diamonds do not affect the diamonds' performance or structural integrity. When set in jewelry, it may also be possible to hide certain inclusion behind mounting hardware such as prongs in a way that renders the defect invisible. However, large clouds can affect a diamond's ability to transmit and scatter light. Large cracks close to or breaking the surface may increase the likelihood of a fracture.

Diamonds are graded by the major societies on a scale ranging from flawless to imperfect.

Color

Jewelers sometimes set diamonds in groups of similar colors.

The Darya-I-Nur Diamond

The finest quality as per color grading is totally colorless, which is graded as D color diamond across the globe, meaning it is absolutely free from any color. The next grade has a very slight trace of color, which can be observed by any expert diamond valuer/grading laboratory. However, when studded in jewellery these very light colored diamonds do not show any color or it is not possible to make out color shades. These are graded as E color or F color diamonds.

Diamonds which show very little traces of color are graded as G or H color diamonds. Slightly colored diamonds are graded as I or J or K color. A diamond can be found in any color in addition to colorless. Some of the colored diamonds, such as pink, are very rare.

A chemically pure and structurally perfect diamond is perfectly transparent with no hue, or *color*. However, in reality most gem-sized natural diamonds are imperfect. The color of a diamond may be affected by chemical impurities and structural defects in the crystal lattice. Depending on the hue and intensity of a diamond's coloration, a diamond's color can either detract from or enhance its value. For example, most white diamonds are discounted in price as a more yellow hue is detectable, while intense pink or blue diamonds (such as the Hope Diamond) can be dramatically more valuable. The Aurora Diamond Collection displays a spectacular array of naturally colored diamonds, which occur in every color of the rainbow.

Most diamonds used as gemstones are basically transparent with little tint, or *white diamonds*. The most common impurity, nitrogen, replaces a small proportion of carbon atoms in a diamond's structure and causes a yellowish to brownish tint. This effect is present in almost all white diamonds; in only the rarest diamonds is the coloration from this effect undetectable. The GIA has developed a rating system for color in white diamonds, from D to Z (with D being "colorless" and Z having a bright yellow coloration), which has been widely adopted in the industry and is universally recognized, superseding several older systems. The GIA system uses a benchmark set of natural diamonds of known color grade, along with standardized and carefully controlled lighting conditions. Diamonds with higher color grades are rarer, in higher demand, and therefore more expensive, than lower color grades. Oddly enough, diamonds graded Z are also rare, and the bright yellow color is also highly valued. Diamonds graded D–F are considered "colorless", G–J are considered "near-colorless", K–M are "slightly colored". N–Y usually appear light yellow or brown.

In contrast to yellow or brown hues, diamonds of other colors are more rare and valuable. While even a pale pink or blue hue may increase the value of a diamond, more intense coloration is usually considered more desirable and commands the highest prices. A variety of impurities and structural imperfections cause different colors in diamonds, including yellow, pink, blue, red, green, brown, and other hues. Diamonds with unusual or intense coloration are sometimes labeled "fancy" in the diamond industry. Intense yellow coloration is considered one of the fancy colors, and is separate from the color grades of white diamonds. Gemologists have developed rating systems for fancy colored diamonds, but they are not in common use because of the relative rarity of such diamonds.

Cut

Diamond cutting is the art and science of creating a gem-quality diamond out of mined rough. The *cut* of a diamond describes the manner in which a diamond has been shaped and polished from its beginning form as a rough stone to its final gem proportions. The cut of a diamond describes the quality of workmanship and the angles to which a diamond is cut. Often diamond cut is confused with "shape".

There are mathematical guidelines for the angles and length ratios at which the diamond is supposed to be cut in order to reflect the maximum amount of light. Round brilliant diamonds, the most common, are guided by these specific guidelines, though fancy cut stones are not able to be as accurately guided by mathematical specifics.

The techniques for cutting diamonds have been developed over hundreds of years, with perhaps the greatest achievements made in 1919 by mathematician and gem enthusiast Marcel Tolkowsky. He developed the round brilliant cut by calculating the ideal shape to return and scatter light when a diamond is viewed from above. The modern round brilliant has 57 facets (polished faces), counting 33 on the *crown* (the top half), and 24 on the *pavilion* (the lower half). The girdle is the thin middle part. The function of the crown is to refract light into various colors and the pavilion's function to reflect light back through the top of the diamond.

Tolkowsky's calculations included some approximations. He calculated the ideal dimensions as:

- Table percentage (corner-to-corner diameter of the table divided by overall diameter) = 53%
- Depth percentage (overall depth divided by overall diameter) = 59.3% (not including adjustments for the culet height and girdle thickness)
- Pavilion angle (angle between the girdle and the pavilion main facets) = 40.75°
- Crown angle (angle between the girdle and the crown's kite facets) = 34.5°
- Pavilion depth (depth of pavilion divided by overall diameter) = 43.1%
- Crown depth (depth of crown divided by overall diameter) = 16.2%

The culet is the tiny point or facet at the bottom of the diamond. This should be a negligible diameter, otherwise light leaks out of the bottom. Tolkowsky's calculations included neither a culet nor a girdle. However, a girdle is required in reality in order to prevent the diamond from easily chipping in the setting. The thick part of the girdle is normally about 1.7% (of the overall diameter) thicker than the thin part of the girdle.

The further the diamond's characteristics are from the Tolkowsky's ideal, the less light will be reflected. However, there is a small range in which the diamond can be considered "ideal". Tolkowsky's calculations can be repeated for a narrow range of pavilion angles. Such calculations show a slightly larger table percentage, and a trade-off between pavilion angle and crown angle.

Today, because of the relative importance of carat weight among buyers, many diamonds are often intentionally cut poorly to increase carat weight. There is a financial premium for a diamond that weighs the desirable 1.0 carat (200 mg), so often the girdle is made thicker or the depth is increased. Neither of these changes makes the diamond *appear* any larger, and both greatly reduce the sparkle of the diamond. (A poorly cut 1.0-carat (200 mg) diamond may have the same diameter and appear as large as a 0.85-carat (170 mg) diamond.) The *depth percentage* is the overall quickest indication of the quality of the cut of a round brilliant. "Ideal" round brilliant diamonds should not have a depth percentage greater than 62.5%. Another quick indication is the *overall diameter*. Typically a round brilliant 1.0-carat (200 mg) diamond should have a diameter of about 6.5 mm (0.26 in). Mathematically, the diameter in millimeters of a round brilliant should approximately equal to 6.5 times the cube root of carat weight, or 11.1 times the cube root of gram weight, or 1.4 times the cube root of point weight.

Shape

Diamonds do not show all of their beauty as rough stones; instead, they must be cut and polished to exhibit the characteristic fire and brilliance that diamond gemstones are known for. Diamonds are cut into a variety of shapes that are generally designed to accentuate these features.

Diamonds which are not cut into a round brilliant shape are known as "fancy cuts." Popular fancy cuts include the *baguette* (French, meaning *rod* or loaf of bread), *marquise*, *princess cut* (square outline), *heart*, *briolette* (a form of the rose cut), and *pear* cuts. Newer cuts that have been introduced into the jewelry industry are the "cushion" "radiant" (similar to princess cuts, but with rounded edges instead of square edges) and Asscher cuts. Many fancy colored diamonds are now being cut according to these new styles. Generally speaking, these "fancy cuts" are not held to the same strict standards as Tolkowsky-derived round brilliants and there are less specific mathematical guidelines of angles which determine a well-cut stone. Cuts are influenced heavily by fashion: the baguette cut—which accentuates a diamond's luster and downplays its fire—was popular during the Art Deco period, whereas the princess cut — which accentuates a diamond's fire rather than its luster — is currently gaining popularity. The princess cut is also popular amongst diamond cutters: of all the cuts, it wastes the least of the original crystal. The past decades have seen the development of new diamond cuts, often based on a modification of an existing cut. Some of these include extra facets. These newly developed cuts are viewed by many as more of an attempt at brand differentiation by diamond sellers, than actual improvements to the state of the art.

Quality

The quality of a diamond's cut is widely considered the most important of the four Cs in determining the beauty of a diamond; indeed, it is commonly acknowledged that a well-cut diamond can appear to be of greater carat weight, and have clarity and color appear to be of better grade than they actually are. The skill with which a diamond is cut determines its ability to reflect and refract light.

In addition to carrying the most importance to a diamond's quality as a gemstone, the cut is also the most difficult to quantitatively judge. A number of factors, including proportion, polish, symmetry, and the relative angles of various facets, are determined by the quality of the cut and can affect the performance of a diamond. A diamond with facets cut only a few degrees out of alignment can result in a poorly performing stone. For a round brilliant cut, there is a balance between "brilliance" and "fire". When a diamond is cut for too much "fire", it looks like a cubic zirconia, which gives off much more "fire" than real diamond. A well-executed round brilliant cut should reflect light upwards and make the diamond appear white when viewed from the top. An inferior cut will produce a stone that appears dark at the center and in extreme cases the setting may be seen through the top of the diamond as shadows.

Several different theories on the "ideal" proportions of a diamond have been and continue to be advocated by various owners of patents on machines to view how well a diamond is cut. These advocate a shift away from grading cut by the use of various angles and proportions toward measuring the performance of a cut stone. A number of specially modified viewers and machines have been developed toward this end. Hearts and Arrows viewers test for the "hearts and arrows" characteristic pattern observable in stones exhibiting high symmetry and particular cut angles. Closely related to Hearts and Arrows viewers is the ASET which tests for light leakage, light return, and proportions. The ASET (and computer simulations of the ASET) are used to test for AGS cut grade. Proponents of these machines argue they help sellers demonstrate the light performance of the diamond in addition to the traditional 4 Cs. Detractors, however, see these machines as marketing tools rather than scientific ones. The GIA has developed a set of criteria for grading the cut of round brilliant stones that is now the standard in the diamond industry and is called Facetware.

Process

The 253-carat (50.6 g) Oppenheimer Diamond, at a 2001 diamond exhibition in Paris. An uncut diamond does not show its prized optical properties.

The process of shaping a rough diamond into a polished gemstone is both an art and a science. The choice of cut is often decided by the original shape of the rough stone, location of the inclusions and flaws to be eliminated, the preservation of the weight, popularity of certain shapes amongst consumers and many other considerations. The round brilliant cut is preferred when the crystal is an octahedron, as often two stones may be cut from one such crystal. Oddly shaped crystals such as macles are more likely to be cut in a *fancy cut*—that is, a cut other than the round brilliant—which the particular crystal shape lends itself to.

Even with modern techniques, the cutting and polishing of a diamond crystal always results in a dramatic loss of weight; rarely is it less than 50%. Sometimes the cutters compromise and accept lesser proportions and symmetry in order to avoid inclusions or to preserve the carat rating. Since the per carat price of diamond shifts around key milestones (such as 1.00 carat (200 mg)), many one-carat diamonds are the result of compromising cut for carat. Some jewelry experts advise consumers to buy a 0.99-carat (198 mg) diamond for its better price or buy a 1.10-carat (220 mg) diamond for its better cut, avoiding a 1.00-carat (200 mg) diamond which is more likely to be a poorly cut stone.

Light Performance

In the gem trade, the term light performance is used to describe how well a polished diamond will return light to the viewer. There are three light properties which are described in relation to light performance: brilliance, fire, and scintillation. Brilliance refers to the white light reflections from the external and internal facet surfaces. Fire refers to the spectral colors which are produced as a result of the diamond dispersing the white light. Scintillation refers to the small flashes of light that are seen when the diamond, light source or the viewer is moved. A diamond that is cut and polished to produce a high level of these qualities is said to be high in light performance.

The setting diamonds are placed in also affect the performance of light through a diamond. The three most commonly used settings are: Prong, Bezel, and Channel. Prong settings are the most popular setting for diamond jewelry. The prong setting consists of four or six 'claws' that cradle the diamond, allowing the maximum amount of light to enter from all angles, allowing the diamonds to appear larger and more brilliant. In bezel settings the diamond or gemstone is completely surrounded by a rim of metal, which can be molded into any shape to accommodate the stone. Used to set earrings, necklaces, bracelets, and rings, bezel settings can have open or closed backs, and generally can be molded to allow a lot of light to pass through. Channel settings set the stones right next to each other with no metal separating them. This setting is mostly used in wedding and anniversary bands. The outer ridge is then worked over the edges of the stones to create a smooth exterior surface. This also protects the girdle area of the stone.

Fluorescence

About a third of all diamonds will glow under ultraviolet light, usually a blue color which may be noticeable under a black light or strong sunlight. According to the GIA, who reviewed a random sample of 26,010 natural diamonds, 65% of the diamonds in the sample had no fluorescence. Of the 35% that did have fluorescence, 97% had blue fluorescence of which 38% had faint blue fluorescence and 62% had fluorescence that ranged from medium to very strong blue. Other colors diamonds can fluoresce are green, yellow, and red but are very rare and are sometimes a combination of the colors such as blue-green or orange. Some diamonds with "very strong" fluorescence can have a "milky" or "oily" look to them, but they are also very rare and are termed "overblues." Their study concluded that with the exception of "overblues" and yellow fluorescent diamonds, fluorescence had little effect on transparency and that the strong and very strong blue fluorescent diamonds on average had better color appearance than non-fluorescent stones. Since blue is a complementary color to yellow and can appear to cancel it out, strong blue fluorescence had especially better color appearance with lower color graded diamonds that have a slight yellowish tint such as I or J color but had little effect on the more colorless D, E and F color grades.

Cleaning

Cleanliness significantly affects a diamond's beauty. A clean diamond is more brilliant and fiery than the same diamond when it is "dirty". Dirt or grease on the top of a diamond reduces its luster. Water, dirt, or grease on the bottom of a diamond interferes with the diamond's brilliance and fire. Even a thin film absorbs some light that could have been reflected to the viewer. Colored dye or smudges can affect the perceived color of a diamond. Historically, some jewelers' stones were misgraded because of smudges on the girdle, or dye on the culet. Current practice is to clean a diamond thoroughly before grading its color.

Maintaining a clean diamond can sometimes be difficult as jewelry settings can obstruct cleaning, and oils, grease, and other hydrophobic materials adhere well to a diamond. Many jewelers use steam cleaners. Some jewelers provide their customers with ammonia-based cleaning kits; ultrasonic cleaners are also popular.

Sapphire

Sapphire is the second hardest crystal that Mother Earth has given us. Second only to the Diamond. Many of the large Sapphire deposits are literally the remains of metamorphic or igneous veins of granite and marble weathered away into stream beds and mountain valleys. Sapphire being stronger than granite or marble survive the weathering and large amounts of the gemstone are simply gathered from gem gravels in Burma, Thailand, Australia, North America, Brazil, Kashmir and Africa.

Sapphire's beauty and enduring strength historically made it quite popular as a talisman among royalty. Legend tells us that medieval kings wore the stone not only to protect themselves from harm, but also to prevent jealousy among their court. A Sapphire worn to the signing of a treaty was believed to ensure a true reconciliation and prevent further bloodshed.

Natural Sapphires

An uncut, rough yellow sapphire found at the Spokane Sapphire Mine near Helena, Montana

Sapphire is one of the two gem-varieties of corundum, the other being ruby(defined as corundum in a shade of red). Although blue is the best-known sapphire color, they occur in other colors, including gray and black, and they can be colorless. A pinkish orange variety of sapphire is called padparadscha.

Significant sapphire deposits are found in Eastern Australia, Thailand, Sri Lanka, China (Shandong), Madagascar, East Africa, and in North Americain a few locations, mostly in Montana. Sapphire and rubies are often found in the same geological setting.

Every sapphire mine produces a wide range of quality – and origin is not a guarantee of quality. For sapphire, Kashmirreceives the highest premium although Burma, Sri Lanka, and Madagascar also produce large quantities of fine quality gems.

The cost of natural sapphires varies depending on their color, clarity, size, cut, and overall quality. For gems of exceptional quality, an independent determination from a respected laboratory such as the GIA, AGL or Gubelin of origin often adds to value.

Blue Sapphire

Teardrop-shaped blue sapphire

Gemstone color can be described in terms of hue, saturation, and tone. Hue is commonly understood as the "color" of the gemstone. Saturation refers to the vividness or brightness of the hue, and tone is the lightness to darkness of the hue. Blue sapphire exists in various mixtures of its primary (blue) and secondary hues, various tonal levels (shades) and at various levels of saturation (vividness).

Blue sapphires are evaluated based upon the purity of their primary hue. Purple, violet, and green are the most common secondary hues found in blue sapphires. Violet and purple can contribute to the overall beauty of the color, while green is considered to be distinctly negative. Blue sapphires with up to 15% violet or purple are generally said to be of fine quality. Gray is the normal saturation modifier or mask found in blue sapphires. Gray reduces the saturation or brightness of the hue, and therefore has a distinctly negative effect.

The color of fine blue sapphires may be described as a vivid medium dark violet to purplish blue where the primary blue hue is at least 85% and the secondary hue no more than 15%, without the least admixture of a green secondary hue or a gray mask.

The 423-carat (84.6 g) Logan sapphire in the National Museum of Natural History, in Washington, D.C., is one of the largest faceted gem-quality blue sapphires in existence.

Dark blue sapphire, probably of Australian origin, showing the brilliant surface luster typical of faceted corundum gemstones.

Sapphires of other Colors

Sapphires in colors other than blue are called "fancy" or "parti colored" sapphires.

Fancy sapphires are often found in yellow, orange, green, brown, purple and violet hues.

Particolored sapphires are those stones which exhibit two or more colors within a single stone. Australia is the largest source of particolored sapphires; they are not commonly used in mainstream jewelry and remain relatively unknown. Particolored sapphires cannot be created synthetically and only occur naturally.

Colorless sapphires have historically been used as diamond substitutes in jewelry.

Pink Sapphires

Pink sapphires occur in shades from light to dark pink, and deepen in color as the quantity of chromium increases. The deeper the pink color, the higher their monetary value. In the United States, a minimum color saturation must be met to be called a ruby, otherwise the stone is referred to as a *pink sapphire*.

Pink sapphire (American terminology, may be classified as rubyoutside of the United States)

Padparadscha

Padparadscha is a delicate, light to medium toned, pink-orange to orange-pink hued corundum, originally found in Sri Lanka, but also found in deposits in Vietnam and parts of East Africa. Padparadscha sapphires are rare; the rarest of all is the totally natural variety, with no sign of artificial treatment.

Faceted padparadscha

The name is derived from the Sanskrit "padma ranga" (padma = lotus; ranga = color), a color akin to the lotus flower (*Nelumbo nucifera*).

Natural padparadscha sapphires often draw higher prices than many of even the finest blue sapphires. Recently, more sapphires of this color have appeared on the market as a result of a new artificial treatment method called "lattice diffusion".

Star Sapphire

A star sapphire is a type of sapphire that exhibits a star-like phenomenon known as asterism; red stones are known as "star rubies". Star sapphires contain intersecting needle-like inclusions following the underlying crystal structure that causes the appearance of a six-rayed "star"-shaped pattern when viewed with a single overhead light source. The inclusion is often the mineral rutile, a mineral composed primarily of titanium dioxide. The stones are cut *en cabochon*, typically with the center of the star near the top of the dome. Occasionally, twelve-rayed stars are found, typically because two different sets of inclusions are found within the same stone, such as a combination of fine needles of rutile with small platelets of hematite; the first results in a whitish star and the second results in a golden-colored star. During crystallisation, the two types of inclusions become preferentially oriented in different directions within the crystal, thereby forming two six-rayed stars that are superimposed upon each other to form a twelve-rayed star. Misshapen stars or 12-rayed stars may also form as a result of twinning. The inclusions can alternatively produce a "cat's eye" effect if the ‹face-up' direction of the cabochon's dome is oriented perpendicular to the crystal's c-axis rather than parallel to it. If the dome is oriented in between these two directions, an 'off-center' star will be visible, offset away from the high point of the dome.

Star sapphire

The Star of Adam is the largest blue star sapphire which weighs 1404.49 carats. The gem was mined in the city of Ratnapura, southern Sri Lanka. The Black Star of Queensland, the second largest gem-quality star sapphire in the world, weighs 733 carats. The Star of India mined in Sri Lanka and weighing 563.4 carats is thought to be the third-largest star sapphire, and is currently on display at the American Museum of Natural History in New York City. The 182-carat Star of Bombay, mined in Sri Lanka and located in the National Museum of Natural History in Washington, D.C., is another example of a large blue star sapphire. The value of a star sapphire depends not only on the weight of the stone, but also the body color, visibility, and intensity of the asterism.

Color Change Sapphire

A rare variety of natural sapphire, known as color-change sapphire, exhibits different color in different light. Color change sapphires are blue in outdoor light and purple under incandescent indoor light, or green to gray-green in daylight and pink to reddish-violet in incandescent light. Color change sapphires come from a variety of locations, including Thailand and Tanzania. The color-change effect is caused by the interaction of the sapphire, which absorbs specific wavelengths of light, and the light-source, whose spectral output varies depending upon the illuminant. Transition-metal impurities in the sapphire, such as chromium and vanadium, are responsible for the color change.

Certain synthetic color-change sapphires have a similar color change to the natural gemstone alexandrite and they are sometimes marketed as "alexandrium" or "synthetic alexandrite". However, the latter term is a misnomer: synthetic color-change sapphires are, technically, not synthetic alexandrites but rather alexandrite *simulants*. This is because genuine alexandrite is a variety of chrysoberyl: not sapphire, but an entirely different mineral.

Source of Color

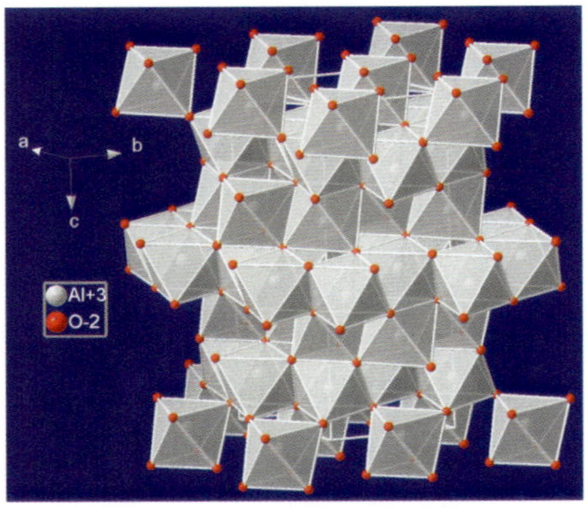

Crystal structure of sapphire

Sapphire ring made circa 1940

Rubies are corundum which contain chromium impurities that absorb yellow-green light and result in deeper ruby redcolor with increasing content. Purple sapphires contain trace amounts of vanadium and come in a variety of shades. Corundum that contains ~0.01% of titanium is colorless. If trace amounts of iron are present, a very pale yellow to green color may be seen. However, if both titanium and iron impurities are present together, and in the correct valence states, the result is a deep-blue color.

Unlike localized ("intra-atomic") absorption of light which causes color for chromium and vanadium impurities, blue color in sapphires comes from intervalence charge transfer, which is the transfer of an electron from one transition-metal ion to another via the conduction or valence band. The iron can take the form Fe^{2+} or Fe^{3+}, while titanium generally takes the form Ti^{4+}. If Fe^{2+} and Ti^{4+} ions are substituted for Al^{3+}, localized areas of charge imbalance are created. An electron transfer from Fe^{2+} and Ti^{4+} can cause a change in the valence state of both. Because of the valence change there is a specific change in energy for the electron, and electromagnetic energy is absorbed. The wavelength of the energy absorbed corresponds to yellow light. When this light is subtracted from incident white light, the complementary color blue results. Sometimes when atomic spacing is different in different directions there is resulting blue-green dichroism.

Intervalence charge transfer is a process that produces a strong colored appearance at a low percentage of impurity. While at least 1% chromium must be present in corundum before the deep red ruby color is seen, sapphire blue is apparent with the presence of only 0.01% of titanium and iron.

Treatments

Sapphires can be treated by several methods to enhance and improve their clarity and color. It is common practice to heat natural sapphires to improve or enhance color. This is done by heating the sapphires in furnaces to temperatures between 500 and 1,800°C (932 and 3,272 °F) for several hours, or by heating in a nitrogen-deficient atmosphere oven for seven days or more. Upon heating, the stone becomes more blue in color, but loses some of the rutile inclusions (silk). When high temperatures are used, the stone loses all silk (inclusions) and it becomes clear under magnification. The inclusions in natural stones are easily seen with a jeweler's loupe. Evidence of sapphire and other gemstones being subjected to heating goes back at least to Roman times. Unheated natural stones are somewhat rare and will often be sold accompanied by a certificate from an independent gemological laboratory attesting to "no evidence of heat treatment".

Yogo sapphire

Yogo sapphires sometimes do not need heat treating because their cornflower blue coloring is uniform and deep, they are generally free of the characteristic inclusions, and they have high uniform clarity. When Intergem Limited began marketing the Yogo in the 1980s as the world's only guaranteed untreated sapphire, heat treatment was not commonly disclosed; by 1982 the heat treatment became a major issue. At that time, 95% of all the world's sapphires were being heated to enhance their natural color. Intergem's marketing of guaranteed untreated Yogos set them against many in the gem industry.

Diffusion treatments are used to add impurities to the sapphire to enhance color. Typically beryllium is diffused into a sapphire under very high heat, just below the melting point of the sapphire. Initially orange sapphires were created, although now the process has been advanced and many colors of sapphire are often treated with beryllium. The colored layer can be removed when stones chip or are repolished or refaceted, depending on the depth of the impurity layer. Treated padparadschas may be very difficult to detect, and many stones are certified by gemological labs (*e.g.*, Gubelin, SSEF, AGTA).

According to United States Federal Trade Commission guidelines, disclosure is required of any mode of enhancement that has a significant effect on the gem's value.

There are several ways of treating sapphire. Heat-treatment in a reducing or oxidising atmosphere (but without the use of any other added impurities) is commonly used to improve the color of sapphires, and this process is sometimes known as "heating only" in the gem trade. In contrast, however, heat treatment combined with the deliberate addition of certain specific impurities (e.g. beryllium, titanium, iron, chromium or nickel, which are absorbed into the crystal structure of the sapphire) is also commonly performed, and this process can be known as "diffusion" in the gem trade. However, despite what the terms "heating only" and "diffusion" might suggest, both of these categories of treatment actually involve diffusion processes.

Mining

Sapphire from Madagascar

Sapphires are mined from alluvial deposits or from primary underground workings. Commercial mining locations for sapphire and ruby include (but are not limited to) the following countries: Afghanistan, Australia, Myanmar/Burma, Cambodia, China, Colombia, India, Kenya, Laos, Madagascar, Malawi, Nepal, Nigeria, Pakistan, Sri Lanka, Tajikistan, Tanzania, Thailand, United States, and Vietnam. Sapphires from different geographic locations may have different appearances or chemical-impurity concentrations, and tend to contain different types of microscopic inclusions. Because of this, sapphires can be divided into three broad categories: classic metamorphic, non-classic metamorphic or magmatic, and classic magmatic.

Sapphires from certain locations, or of certain categories, may be more commercially appealing than others,particularly classic metamorphic sapphires from Kashmir, Burma, or Sri Lanka that have not been subjected to heat-treatment.

The Logan sapphire, the Star of India, and the Star of Bombay originate from Sri Lankan mines. Madagascar is the world leader in sapphire production (as of 2007) specifically its deposits in and around the town of Ilakaka. Prior to the opening of the Ilakaka mines, Australia was the largest producer of sapphires (such as in 1987). In 1991 a new source of sapphires was discovered in Andranondambo, southern Madagascar. That area has been exploited for its sapphires started in 1993, but it was practically abandoned just a few years later—because of the difficulties in recovering sapphires in their bedrock.

In North America, sapphires have been mined mostly from deposits in Montana: fancies along the Missouri River near Helena, Montana, Dry Cottonwood Creek near Deer Lodge, Montana, and Rock Creek near Philipsburg, Montana. Fine blue Yogo sapphires are found at Yogo Gulch west of Lewistown, Montana. A few gem-grade sapphires and rubies have also been found in the area of Franklin, North Carolina.

The sapphire deposits of Kashmir are well known in the gem industry, although their peak production took place in a relatively short period at the end of the nineteenth and early twentieth centuries. They have a superior cornflower blue hue to them with a mysterious and almost sleepy quality, described by some gem enthusiasts as 'blue velvet". Kashmir-origin contributes meaningfully to the value of a sapphire, and most corundum of Kashmir origin can be readily identified by its characteristic silky appearance and exceptional hue. The unique blue appears lustrous under any kind of light, unlike non-Kashmir sapphires which may appear purplish or grayish in comparison. Sotheby's has been in the forefront overseeing record-breaking sales of Kashmir sapphires worldwide. In October 2014, Sotheby's Hong Kong achieved consecutive per-carat price records for Kashmir sapphires - first with the 12.00 carat Cartier sapphire ring at US$193,975 per carat, then with a 17.16 carat sapphire at US$236,404, and again in June 2015 when the per-carat auction record was set at US$240,205. At present, the world record price-per-carat for sapphire at auction is held by a sapphire from Kashmir in a ring, which sold in October 2015 for approximately US$242,000 per carat (HK$52,280,000 in total, including buyer's premium, or more than US$6.74 million).

Synthetic Sapphire

Synthetic sapphire

Synthetic star sapphire

In 1902, the French chemist Auguste Verneuil developed a process for producing synthetic sapphire crystals. In the Verneuil process, named after him, fine alumina powder is added to an oxyhydrogen flame, and this is directed downward against a mantle. The alumina in the flame is slowly deposited, creating a teardrop shaped "boule" of sapphire material. Chemical dopants can be added to create artificial versions of the ruby, and all the other natural colors of sapphire, and in addition, other colors never seen in geological samples. Artificial sapphire material is identical to natural sapphire, except it can be made without the flaws that are found in natural stones. The disadvantage of Verneuil process is that the grown crystals have high internal strains. Many methods of manufacturing sapphire today are variations of the Czochralski process, which was invented in 1916 by Polish chemist Jan Czochralski. In this process, a tiny sapphire seed crystal is dipped into a crucible made of the precious metal iridium or molybdenum, containing molten alumina, and then slowly withdrawn upward at a rate of 1 to 100 mm per hour. The alumina crystallizes on the end, creating long carrot-shaped boules of large size up to 200 kg in mass.

Synthetic sapphire is also produced industrially from agglomerated aluminium oxide, sintered and fused (such as by hot isostatic pressing) in an inert atmosphere, yielding a transparent but slightly porous polycrystalline product.

In 2003, the world's production of synthetic sapphire was 250 tons (1.25×10^9 carats), mostly by the United States and Russia. The availability of cheap synthetic sapphire unlocked many industrial uses for this unique material.

The first laser was made with a rod of synthetic ruby. Titanium-sapphire lasers are popular due to their relatively rare capacity to be tuned to various wavelengths in the red and near-infrared region of the electromagnetic spectrum. They can also be easily mode-locked. In these lasers a synthetically produced sapphire crystal with chromium or titaniumimpurities is irradiated with intense light from a special lamp, or another laser, to create stimulated emission.

Common Applications

Windows

Cermax xenon arc lamp with synthetic sapphire output window

Synthetic sapphire—sometimes referred to as *sapphire glass*—is commonly used as a window material, because it is both highly transparent to wavelengths of light between 150 nm (UV) and 5500 nm (IR) (the visible spectrum extends about 380 nm to 750 nm), and extraordinarily scratch-resistant.

The key benefits of sapphire windows are:

- Very wide optical transmission band from UV to near-infrared, (0.15–5.5 μm).

- Significantly stronger than other optical materials or standard glass windows.
- Highly resistant to scratching and abrasion (9 on the Mohs scale of mineral hardness scale, the 3rd hardest natural substance next to moissanite and diamonds).
- Extremely high melting temperature (2030° C).

Single crystal sapphire boule grown by the Kyropoulos method.
Approximately 200 mm diameter and approximately 30 kg.

Some sapphire-glass windows are made from pure sapphire boules that have been grown in a specific crystal orientation, typically along the optical axis, the c-axis, for minimum birefringence for the application.

The boules are sliced up into the desired window thickness and finally polished to the desired surface finish. Sapphire optical windows can be polished to a wide range of surface finishes due to its crystal structure and its hardness. The surface finishes of optical windows are normally called out by the scratch-dig specifications in accordance with the globally adopted MIL-O-13830 specification.

The sapphire windows are used in both high pressure and vacuum chambers for spectroscopy, crystals in various watches, and windows in grocery store barcode scanners since the material's exceptional hardness and toughness makes it very resistant to scratching.

It is used for end windows on some high-powered laser tubes as its wide-band transparency and thermal conductivity allow it to handle very high power densities in the infra-red or UV spectrum without degrading due to heating.

Along with zirconia and aluminium oxynitride, synthetic sapphire is used for shatter resistant windows in armored vehicles and various military body armor suits, in association with composites.

One type of xenon arc lamp – originally called the "Cermax" and now known generically as the "ceramic body xenon lamp" – uses sapphire crystal output windows. This product tolerates higher thermal loads and thus higher output powers when compared with conventional Xe lamps with pure silica window.

Use as Substrate for Semiconducting Circuits

Thin sapphire wafers were the first successful use of an insulating substrate upon which to deposit silicon to make the integrated circuits known as silicon on sapphire or "SOS"; now other substrates can also be used for the class of circuits known more generally as silicon on insulator. Besides its excellent electrical insulating properties, sapphire has high thermal conductivity. CMOS chips on sapphire are especially useful for high-power radio-frequency (RF) applications such as those found in cellular telephones, public-safety band radios, and satellite communication systems. "SOS" also allows for the monolithic integration of both digital and analog circuitry all on one IC chip, and the construction of extremely low power circuits.

In one process, after single crystal sapphire boules are grown, they are core-drilled into cylindrical rods, and wafers are then sliced from these cores.

Wafers of single-crystal sapphire are also used in the semiconductor industry as substrates for the growth of devices based on gallium nitride (GaN). The use of sapphire significantly reduces the cost, because it has about one-seventh the cost of germanium. Gallium nitride on sapphire is commonly used in blue light-emitting diodes (LEDs).

Use for Endoprostheses

Monocrystalline sapphire is fairly biocompatible and the exceptionally low wear of sapphire–metal pairs has led to the introduction (in Ukraine) of sapphire monocrystals for hip joint endoprostheses.

Ruby

A ruby is a pink to blood-red colored gemstone, a variety of the mineral corundum (aluminium oxide). Other varieties of gem-quality corundum are called sapphires. Ruby is one of the traditional cardinal gems, together with amethyst, sapphire, emerald, and diamond. The word *ruby* comes from *ruber*, Latin for red. The color of a ruby is due to the element chromium.

The quality of a ruby is determined by its color, cut, and clarity, which, along with carat weight, affect its value. The brightest and most valuable shade of red called blood-red or pigeon blood, commands a large premium over other rubies of similar quality. After color follows clarity: similar to diamonds, a clear stone will command a premium, but a ruby without any needle-like rutile inclusions may indicate that the stone has been treated. Ruby is the traditional birthstone for July and is usually more pink than garnet, although some rhodolite garnets have a similar pinkish hue to most rubies. The world's most valuable ruby is the Sunrise Ruby.

Physical Properties

Rubies have a hardness of 9.0 on the Mohs scale of mineral hardness. Among the natural gems only moissanite and diamond are harder, with diamond having a Mohs hardness of 10.0 and moissanite falling somewhere in between corundum (ruby) and diamond in hardness. Sapphire, ruby, and pure corundum are α-alumina, the most stable form of Al_2O_3, in which 3 electrons leave each aluminum ion to join the regular octahedral group of six nearby O^{2-} ions; in pure corundum this leaves all of the aluminum ions with a very stable configuration of no unpaired electrons or unfilled energy levels, and the crystal is perfectly colorless.

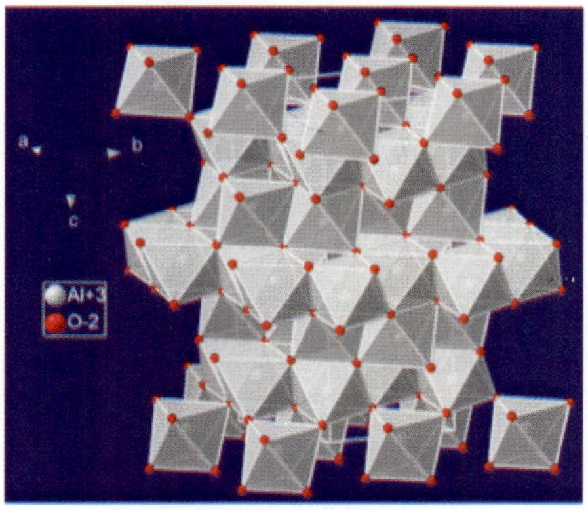

Crystal structure of rubies

When a chromium atom replaces an occasional aluminum atom, it too loses 3 electrons to become a chromium^{3+} ion to maintain the charge balance of the Al_2O_3 crystal. However, the Cr^{3+} ions are larger and have electron orbitals in different directions than aluminum. The octahedral arrangement of the O^{2-} ions is distorted, and the energy levels of the different orbitals of those Cr^{3+} ions are slightly altered because of the directions to the O^{2-} ions. Those energy differences correspond to absorption in the ultraviolet, violet, and yellow-green regions of the spectrum.

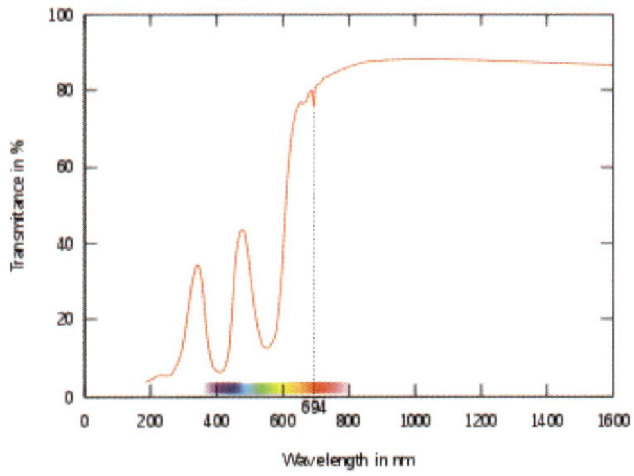

Transmittance of ruby in optical and near-IR spectra. Note the two broad violet and yellow-green absorption bands and one narrow absorption band at the wavelength of 694 nm, which is the wavelength of the ruby laser.

If one percent of the aluminum ions are replaced by chromium in ruby, the yellow-green absorption results in a red color for the gem. Additionally, absorption at any of the above wavelengths stimulates fluorescent emission of 694-nanometer-wavelength red light, which adds to its red color and perceived luster.

After absorbing short-wavelength light, there is a short interval of time when the crystal lattice of ruby is in an excited state before fluorescence occurs. If 694-nanometer photons pass through the crystal during that time, they can stimulate more fluorescent photons to be emitted in-phase with them, thus strengthening the intensity of that red light. By arranging mirrors or other means to pass emitted light repeatedly through the crystal, a ruby laser in this way produces a very high intensity of coherent red light.

All natural rubies have imperfections in them, including color impurities and inclusions of rutile needles known as "silk". Gemologists use these needle inclusions found in natural rubies to distinguish them from synthetics, simulants, or substitutes. Usually, the rough stone is heated before cutting. These days, almost all rubies are treated in some form, with heat treatment being the most common practice. Untreated rubies of high quality command a large premium.

Some rubies show a three-point or six-point asterism or "star". These rubies are cut into cabochons to display the effect properly. Asterisms are best visible with a single-light source and move across the stone as the light moves or the stone is rotated. Such effects occur when light is reflected off the "silk" (the structurally oriented rutile needle inclusions) in a certain way. This is one example where inclusions increase the value of a gemstone. Furthermore, rubies can show color changes—though this occurs very rarely—as well as chatoyancy or the "cat's eye" effect.

Ruby vs. Pink Sapphire

Generally, gemstone-quality corundum in all shades of red, including pink, are called rubies. However, in the United States, a minimum color saturation must be met to be called a ruby; otherwise, the stone will be called a pink sapphire.Drawing a distinction between rubies and pink sapphires is relatively new, having arisen sometime in the 20th century. Often, the distinction between ruby and pink sapphire is not clear and can be debated. As a result of the difficulty and subjectiveness of such distinctions, trade organizations such as the International Colored Gemstone Association (ICGA) have adopted the broader definition for ruby which encompasses its lighter shades, including pink.

Natural Occurrence

The Mogok Valley in Upper Myanmar (Burma) was for centuries the world's main source for rubies. That region has produced some exceptional rubies, however in recent years few good rubies have been found. In central Myanmar, the area of Mong Hsu began producing rubies during the 1990s and rapidly became the world's main ruby mining area. The most recently found ruby deposit in Myanmar is in Namya (Namyazeik) located in the northern state of Kachin.

Historically, rubies have also been mined in Thailand, in the Pailin and Samlout District of Cambodia, as well as in Afghanistan, Australia, Brazil, Colombia, India, Namibia, Japan, and Scotland; after the Second World War ruby deposits were found in Madagascar, Nepal, Pakistan, Tajikistan, Tanzania, and Vietnam. In Sri Lanka, lighter shades of rubies (often "pink sapphires") are more commonly found. The Republic of Macedonia is the only country in mainland Europe to have naturally occurring rubies. They can mainly be found around the city of Prilep. Macedonian rubies have a unique raspberry color. The ruby is also included on the Macedonian coat of arms. A few rubies have been found in the U.S. states of Montana, North Carolina, South Carolina and Wyoming.

Spinel, another red gemstone, is sometimes found along with rubies in the same gem gravel or marble. Red spinels may be mistaken for rubies by those lacking experience with gems. However, the finest red spinels can have values approaching that of an average ruby.

Factors Affecting Value

Rubies, as with other gemstones, are graded using criteria known as the four Cs, namely color, cut, clarity and carat weight. Rubies are also evaluated on the basis of their geographic origin.

Color: In the evaluation of colored gemstones, color is the most important factor. Color divides into three components: *hue, saturation* and *tone*. Hue refers to color as we normally use the term. Transparent gemstones occur in the *pure spectral hues* of red, orange, yellow, green, blue, violet. In nature, there are rarely pure hues, so when speaking of the hue of a gemstone, we speak of primary and secondary and sometimes tertiary hues. Ruby is defined to be red. All other hues of the gem species corundum are called sapphire. Ruby may exhibit a range of secondary hues, including orange, purple, violet, and pink.

Natural ruby with inclusions

A cut pink ruby

Purple rubies

A naturally occurring ruby crystal

Victorian ruby and diamond brooch

The finest ruby is described as being a vivid medium-dark toned red. Secondary hues add an additional complication. Pink, orange, and purple are the normal secondary hues in ruby. Of the three, purple is preferred because it reinforces the red, making it appear richer. Purple also occupies a position on the color wheel halfway between red and blue. When a purplish-red ruby is set in yellow, the yellow neutralizes its complement blue, leaving the stone appearing to be pure red in the setting.

Treatments and Enhancements

Improving the quality of gemstones by treating them is common practice. Some treatments are used in almost all cases and are therefore considered acceptable. During the late 1990s, a large supply of low-cost materials caused a sudden surge in supply of heat-treated rubies, leading to a downward pressure on ruby prices.

Improvements used include color alteration, improving transparency by dissolving rutile inclusions, healing of fractures (cracks) or even completely filling them.

The most common treatment is the application of heat. Most rubies at the lower end of the market are heat treated to improve color, remove *purple tinge*, blue patches, and silk. These heat treatments typically occur around temperatures of 1800°C (3300 °F). Some rubies undergo a process of low tube heat, when the stone is heated over charcoal of a temperature of about 1300°C (2400 °F) for 20 to 30 minutes. The silk is partially broken, and the color is improved.

Another treatment, which has become more frequent in recent years, is lead glass filling. Filling the fractures inside the ruby with lead glass (or a similar material) dramatically improves the transparency of the stone, making previously unsuitable rubies fit for applications in jewelry. The process is done in four steps:

1. The rough stones are pre-polished to eradicate all surface impurities that may affect the process.
2. The rough is cleaned with hydrogen fluoride.
3. The first heating process during which no fillers are added. The heating process eradicates impurities inside the fractures. Although this can be done at temperatures up to 1400 °C (2500 °F) it most likely occurs at a temperature of around 900 °C (1600 °F) since the rutile silk is still intact.
4. The second heating process in an electrical oven with different chemical additives. Different solutions and mixes have shown to be successful, however mostly lead-containing glass-powder is used at present. The ruby is dipped into oils, then covered with powder, embedded on a tile and placed in the oven where it is heated at around 900 °C (1600 °F) for one hour in an oxidizing atmosphere. The orange colored powder transforms upon heating into a transparent to yellow-colored paste, which fills all fractures. After cooling the color of the paste is fully transparent and dramatically improves the overall transparency of the ruby.

If a color needs to be added, the glass powder can be "enhanced" with copper or other metal oxides as well as elements such as sodium, calcium, potassium etc.

The second heating process can be repeated three to four times, even applying different mixtures. When jewelry containing rubies is heated (for repairs) it should not be coated with boracic acid or any other substance, as this can etch the surface; it does not have to be "protected" like a diamond.

The treatment can identify by noting bubbles in cavities and fractures using a 10 x loupe.

Synthetic and Imitation Rubies

Artificial ruby under a normal light (top) and under a green laser light (bottom). Red light is emitted

In 1837, Gaudin made the first synthetic rubies by fusing potash alum at a high temperature with a little chromium as a pigment. In 1847, Ebelmen made white sapphire by fusing alumina in boric acid. In 1877, Frenic and Freil made crystal corundum from which small stones could be cut. Frimy and Auguste Verneuil manufactured artificial ruby by fusing BaF_2 and Al_2O_3 with a little chromium at red heat. In 1903, Verneuil announced he could produce synthetic rubies on a commercial scale using this flame fusion process. By 1910, Verneuil's laboratory had expanded into a 30 furnace production facility, with annual gemstone production having reached 1,000 kilograms (2,000 lb) in 1907.

Other processes in which synthetic rubies can be produced are through Czochralski's pulling process, flux process, and the hydrothermal process. Most synthetic rubies originate from flame fusion, due to the low costs involved. Synthetic rubies may have no imperfections visible to the naked eye but magnification may reveal curves, striae and gas bubbles. The fewer the number and the less obvious the imperfections, the more valuable the ruby is; unless there are no imperfections (i.e., a perfect ruby), in which case it will be suspected of being artificial. Dopants are added to some manufactured rubies so they can be identified as synthetic, but most need gemological testing to determine their origin.

Synthetic rubies have technological uses as well as gemological ones. Rods of synthetic ruby are used to make ruby lasersand masers. The first working laser was made by Theodore H. Maiman in 1960. Maiman used a solid-state light-pumped synthetic ruby to produce red laser light at a wavelength of 694 nanometers (nm). Ruby lasers are still in use. Rubies are also used in applications where high hardness is required such as at wear exposed locations in modern mechanical clockworks, or as scanning probe tips in a coordinate measuring machine.

Imitation rubies are also marketed. Red spinels, red garnets, and colored glass have been falsely claimed to be rubies. Imitations go back to Roman times and already in the 17th century techniques were developed to color foil red—by burning scarlet wool in the bottom part of the furnace—which was then placed under the imitation stone. Trade terms such as balas ruby for red spinel and rubellite for red tourmaline can mislead unsuspecting buyers. Such terms are therefore discouraged from use by many gemological associations such as the Laboratory Manual Harmonisation Committee (LMHC).

Emerald

Another one of the five cardinal gemstones, the emerald, is a variety of the mineral beryl. The main component of the mineral is $Be_3Al_2(SiO_3)_6$. Trace amounts of chromium and occasionally, vanadium is also present in this mineral. Emeralds are characterized by their green color. To be labeled as a top gem, an emerald must have a pure verdant green hue and also exhibit a high level of transparency. In ancient times, Egypt, India, and Austria were important emerald mining sites. Today, however, Colombia is regarded as the world's top producer of this gemstone. The Chalk Emerald, the Mogul Mughal Emerald, etc., are some of the world famous emeralds.

Identifying Emerald

Almost all natural emeralds contain distinct characteristic inclusions and almost all are treated with oil or resin to fill tiny fissures and cracks. It is the presence of these flaws and oil which makes it quite easy to identify and distinguish natural emeralds amongst other similar gemstones.

Artificial light will expose and amplify inclusions and fractures that prove the stone to be a natural emerald. Emerald color is owed to trace amounts of chromium and vanadium, and color is best admired under natural daylight. One of the easiest methods to identify green emeralds is to test for specific gravity (density) and hardness. Like all forms of beryl, emerald is harder than apatite, quartz and feldspar, but is slightly softer than spinel, topaz and sapphire. However, emerald is generally more fragile than other beryl, owing to its naturally included state.

Emerald Origin and Gemstone Sources

Emerald mining is almost exclusively done from host rocks, where the emerald has grown into small veins or on walls of cavities.

Colombia is the international center of emerald mining. The Muzo Mine, northwest of Bogotá, produces fine-quality stones of a deep green color. The Chivor Mine, northeast of Bogotá, is another important deposit, and other locations promise additional successful mining results.

Brazil has various deposits in Bahia, Goias and Minas Gerais. The stones are lighter than the Colombian ones; mostly yellow-green and are often free of inclusions. Brazil also supplies rare emerald cat's eyes and extremely rare emeralds with a six-ray star.

South African deposits are concentrated in the Northern Transvaal. But only five percent of the stones found in the Cobra and Somerset Mines are of good quality. Most stones are light or muddy and are only suitable for cabochon cuts.

Zimbabwe has several deposits of emerald; the most important is the Sandawana Mine in the south. The crystals are small, but of very good quality.

Russia has deposits in the Ural, north of Sverdlovsk. Good quality emeralds are rare; most stones are light or muddy and only suitable for cabochons.

Further deposits are in Afghanistan, Australia (New South Wales and Western Australia), Ghana, India, Madagascar, Malawi, Mozambique, Namibia, Nigeria, Pakistan, Tanzania, Zambia and the United States (North Carolina).

Emerald Color

With emeralds, even more so than other colored gems, it is the color that is the chief determinant of value. By definition, emeralds are a medium to darker green to blue-green and sometimes slightly yellowish-green beryl. Emeralds owe their fine green color to traces of chromium and vanadium impurities. The most popular and valuable color is a slightly bluish green in a medium dark tone with strong to vivid saturation, however, too much blue can decrease the value.

The term "Colombian emeralds" is often used to describe vivid, slightly bluish-green stones of a medium to medium dark color, regardless of their geographic origin. Emeralds of a lighter color are sometimes called "Brazilian emeralds", even if they were mined in Africa.

Emerald Clarity and Luster

Emeralds have a vitreous (glass-like) luster when cut and polished. Although their clarity can occur from translucent to opaque, transparent specimens are most desirable and are much more valuable. Clarity is important, but inclusions are tolerated more in emeralds than virtually any other gem. Unlike other beryl gems, emeralds often contain inclusions and other flaws. These flaws are not looked upon as negative attributes for emeralds, as they would be for other gemstones. Indeed, these flaws are considered part of the character of the stone and are used to assure the purchaser of a natural stone.

Emerald clarity is graded by eye, unlike diamond, where 10x loupe magnification is used to grade clarity. If an emerald has no inclusions that are visible to the naked eye, it is considered flawless. "Eye-clean" stones will command the highest prices, especially those with ideal color grade.

Emerald Cut and Shape

Emeralds are most often cut in a special cut designed just for this gem; "the emerald-cut". The emerald cut is a step-cut or trap-cut featuring a rectangular or square shape with truncated corners. This cutting style maximizes the beauty and color of the stone, whilst protecting it from mechanical strain and internal stress. Emeralds are also cut into a variety of other traditional shapes such as pear, oval and round. Lower grade materials are often cut en cabochon or into beads. Highly transparent and clear materials are sometimes cut in brilliant-style.

Emerald Treatment

Oiling is a common emerald treatment and in most cases, it is done right at the mining location. The term ‹oiling' refers to the practice of immersing emeralds in a colorless oil or resin (most often cedar oil) in order to enhance color, clarity and stability. This is also often done using a vacuum chamber to assist penetration. Non-standard treatments go beyond this by using colored oils and epoxy-like resins.

These treatments dramatically improve the appearance of the gems, but necessitate special care in cleaning and setting. Steam cleaners, solvents and ultrasonic cleaners can remove the oils, making inclusions that were barely visible stand out in sharp relief. Since emeralds can be restored through re-oiling, the damage is considered to be only temporary.

Synthetic emeralds have been available since 1848. In the 1950s, synthesizing methods became commercialized and excellent quality synthetics have been made available since. The term 'Chatham emeralds' is often used for lab-grown emeralds. There are also numerous emerald doublets available. In many cases, a natural upper-half is cemented to a glass or synthetic pavilion using an emerald-green paste. With the visibility of emerald flaws from the crown down, once these are set into jewelry, they are often sold as natural emeralds.

Emerald Gemological Properties

Species:	Emerald - beryl
Chemical Formula:	$Al_2Be_3Si_6O_{18}$ - Aluminum beryllium silicate
Crystal Structure:	Hexagonal; hexagonal prisms
Color:	Emerald green to green to slightly bluish or yellowish-green
Hardness:	7.5 to 8 on the Mohs scale

Refractive Index:	1.565 to 1.602
Density:	2.67 to 2.78
Cleavage:	Indistinct
Transparency:	Transparent to opaque
Double Refraction or Birefringence:	-0.006
Luster:	Vitreous
Fluorescence:	Usually none

Synthetic Emerald

Emerald showing its hexagonal structure

Both hydrothermal and *flux-growth* synthetics have been produced, and a method has been developed for producing an emerald overgrowth on colorless beryl. The first commercially successful emerald synthesis process was that of Carroll Chatham, likely involving a lithium vanadate flux process, as Chatham's emeralds do not have any water and contain traces of vanadate, molybdenum and vanadium. The other large producer of flux emeralds was Pierre Gilson Sr., whose products have been on the market since 1964. Gilson's emeralds are usually grown on natural colorless beryl seeds, which are coated on both sides. Growth occurs at the rate of 1 mm per month, a typical seven-month growth run produces emerald crystals 7 mm thick.

Hydrothermal synthetic emeralds have been attributed to IG Farben, Nacken, Tairus, and others, but the first satisfactory commercial product was that of Johann Lechleitner of Innsbruck, Austria, which appeared on the market in the 1960s. These stones were initially sold under the names "Emerita" and "Symeralds", and they were grown as a thin layer of emerald on top of natural colorless beryl stones. Later, from 1965 to 1970, the Linde Division of Union Carbide produced completely synthetic emeralds by hydrothermal synthesis. According to their patents (attributable to E.M. Flanigen), acidic conditions are essential to prevent the chromium (which is used as the colorant) from precipitating. Also, it is important that the silicon-containing nutrient be kept away from the other ingredients to prevent nucleation and confine growth to the seed crystals. Growth occurs by a diffusion-reaction process, assisted by convection. The largest producer of hydrothermal emeralds today is Tairus, which has succeeded in synthesizing emeralds with chemical composition similar to emeralds in alkaline deposits in Colombia, and whose products are thus known as "Colombian created emeralds" or "Tairus created emeralds". Luminescence in ultraviolet light is considered a supplementary test when making a natural versus synthetic determination, as many, but not all, natural emeralds are inert to ultraviolet light. Many synthetics are also UV inert.

Synthetic emeralds are often referred to as "created", as their chemical and gemological composition is the same as their natural counterparts. The U.S. Federal Trade Commission (FTC) has very strict regulations as to what can and what cannot be called a "synthetic" stone. The FTC says: "(c) It is unfair or deceptive to use the word "laboratory-grown," "laboratory-created," "manufacturer name-created," or "synthetic" with the name of any natural stone to describe any industry product unless such industry product has essentially the same optical, physical, and chemical properties as the stone named."

Amethyst

Its color is as unique as it is seductive, though in fact this gemstone of all gemstones is said to protect its wearer against seduction.

The amethyst is extravagance in violet. For many thousands of years, the most striking representative of the quartz family has been a jewel coveted by princes both ecclesiastical and secular. Moses described it as a symbol of the Spirit of God in the official robes of the High Priest of the Jews, and the Russian Empress Catherine the Great sent thousands of miners into the Urals to look for it. In popular belief, the amethyst offers protection against drunkenness - for the Greek words 'amethystos' mean 'not intoxicated' in translation. A more apt stone for the month of February, particularly if there is to be plenty going on in the way of carnival celebrations, could thus hardly be wished for.

A large number of further miraculous powers are attributed to the amethyst in all sorts of cultures. It was said to protect crops against tempests and locusts, bring good fortune in war and in the hunt, drive out evil spirits and inspire the intellect. A little study of the works of Pliny will reveal that this gemstone, if worn round the neck on a cord made from dog's hair, affords protection against snakebite. Later, Hieronymus even reported that eagles placed an amethyst in their nest in order to protect their young from the selfsame danger. Apart from these powers, gemstone therapists say that the amethyst has a sobering and cleansing effect.

Amethyst has also been said to quell excessive stomach acid and, served to combat insect bites and beautify the skin. But the amethyst not only had a firm niche in medicine; it was also esteemed as a stone of friendship. And since it was thought to put the wearer in a chaste frame of mind and symbolize trust and piety, the amethyst came to occupy a very prominent position in the ornaments of the Catholic clergy over the centuries. It was the stone of bishops and cardinals; we find it in prelates' crosses and in the so-called Papal Ring (Italian, 15th century) in the Jewelry Museum in Pforzheim.

However, the most beautiful of all crystal quartzes also posed one or two riddles for the scientists, and indeed they still haven't been completely solved to this day. The amethyst has its hardness, its moderate refraction and its weight in common with the other quartzes, but the crystal structure is different, and it is most unconventional. The construction is stratified, as a result of which areas and lamellae of varying color intensity often come about. This explains why there are relatively few large cut amethysts of an evenly distributed dark color, in spite of its having been found so abundantly in all parts of the world. It is only in the last few years that scientists have been fairly certain of having found the real cause of the color. It is now attributed to certain iron constituents in connection with natural radioactive radiation.

One thing that has been known for a long time, on the other hand, is the fact that the amethyst changes its color on being heated. Smoky stones are transformed at temperatures of as little as 250 degrees to a shining yellow to brownish-red, whilst clear ones, i.e. those with a high degree of transparency, become yellow or colorless at 400 degrees. Now and then Nature gives us a surprise by having created bicolored stones, like the ones recently found in Bolivia in the form of causticised crystal nuggets. This variety is known as ametrine, for in its formation certain energy states of iron introduce violet areas to the yellow citrine. At best, flat jewelry with a three-spoked star can be cut from it. However, the highlight for esoterics is that the energy fields can in fact be made visible in polarised light. The Henn Brothers of Idar-Oberstein even supply the photos to go with it.

Some amethysts pale almost to colourlessness in daylight. The reason for this has not yet been discovered, but it is possible to re-colour them by means of radium radiation. The fact that these stones can lose their colour makes it obvious that amethyst jewellery should not be worn while sunbathing, in a solarium or in a discotheque with black light. Sudden changes of temperature can also be harmful to the stone.

Asterism

Asterism" is a term, which is most commonly used in gemology. It refers a kind of special optic phenomenon that has been seen in several gems.

Special optic phenomena refer uncommon and excellent optic properties exhibit by some rare gemstone. The phenomena are particular property of transmitting or reflecting light. The structure or inclusions of a mineral in gems is the reason of this property. The beauty and value of gems depend on their optical properties.

"Asterism"-one of the important optical phenomena has been seen in sapphires, rubies, rose quartz and some other gems. The term "Asterism" came from the Greek word "aster", which means star. From the name we can guess "asterism" is star-shaped light effect across the surface of gems.

Special Optical Phenomenon

"Asterism" is an exclusive light effect (which we called special optical phenomenon) of some crystal. When light pass through these crystals, its exhibit a star-shaped figure, which called "asterism". This figure is produced by reflected or transmitted light.

Asterism in Gems

Gems, which show an illuminated star, just bellow the surface of crystal called star gems. The most popular star gems are star ruby and star sapphire. Other gems, which also possess the special optic phenomenon "asterism", are rose quartz, garnet, spinel, Chrysoberyl and Diopside. There are a few specimen of rock crystal, which has the phenomenon.

The star in the star gems may have four to twelve rays depending on the crystal system of the gems. But the six- rayed star effect is most common. Usually, six-rayed star show by the star rubies and star sapphires. Sometimes these star gems exhibit the twelve-rayed star effect. But the twelve-rayed star effect in corundum is occasional. Rose quartz exhibits six-rayed star. Some garnets and spinel commonly have the star effect with four rays, but these star gems can show six-rayed star effect if cut in a distinct way. Some crystals like Chrysoberyl and Diopside only exhibit four-rayed star effect.

"Asterism" is caused by the dense inclusions of tiny fibres of foreign minerals in gems. Generally, these inclusions are the reason of the reflection of light and form a wavy star like formation, which moves around when the mineral is rotated. All the inclusions in the gems are formed via exsolution.

The inclusions must be long, very thin needle like which called as rutile. "*Asterism*" is produced when needles intersect each other in multiple groups. Small needles of rutile (titanium oxide), sometimes called "silk" are oriented inside the minerals in the direction of the crystal growth system. There is a relationship between the gem's growth and the number of rays produced by the reflected light.

There must be three or more chatoyant bands inside the gems to produce "asterism". The arrangement of needles in a hexagonal pattern produced six-rayed "asterism".

Exsolved inclusions form a pattern in the host mineral. The pattern may vary from mineral to mineral, but in case of specific mineral the pattern remain same. In case of corundum, we can see rutile is exsolved in three directions crossing at 60 degree or 120 degree in the basal and produce six or twelve-rayed star. When the rutiles crossed at 90 degree, it produces a four-rayed star. Star gems are translucent to opaque.The rays of star should cross at the top of these gems and be distinct.

Star Sapphire

The sharpness and the size of "asterism" in gems, which identified as phenomenon, depend on the quality and quantity of rutile needles. Small to smaller size of rutile needles help "asterism" to be sharper. When the needles are visible with open eyes as inclusions, the gems turned into a low translucent stone and lack "asterism".

Large quantities of rutile needles in a crystal produce phenomenal "aster". If the quantities of rutiles are small, "Asterism" is visible in only a small area. These gems are not recognized as phenomenal gems.The clarity of these gems can be improve by dissolving some rutile needles with heat treatment. Heat treatment allows various elements to be incorporated into crystal. Then rutile needles form by very gradual cooling system and induce "asterism".

Inclusions are not present in all star gems. Some specimens leave holes or empty areas in the host. Reflection creates from fibres or fibrous cavity in the crystal.

Types of Asterism

There are two types of "asterism", like Epiasterism and Diasterism. Epiasterism is creates by the reflection of light on parallel-arranged inclusions inside the crystal. Diasterism is creates by the transmitting light through the stone, mostly seen in rose quartz. To see this effect a lighted background is essential.

To display the star effect in best way a gem is cut as cabochon with the girdle parallel to the needles. The domed stones are round or oval shaped.

The special optical phenomenon "asterism" is most visible in direct, single pointed light source, like an incandescent bulb, fibre optic light, and penlight or in sunlight.

The distinctiveness of the star and degree of strait ness of each ray help to judge the quality and value of star gems.

Chatoyancy

Chatoyance is an optical phenomenon in which a band of reflected light, known as a "cat's-eye," moves just beneath the surface of a cabochon-cut gemstone. Chrysoberyl and tiger's-eye are two of the best-known gem materials that exhibit this phenomenon. Excellent specimens of chrysoberyl exhibit the finest chatoyance, and tiger's-eye is the chatoyant gem most widely used in jewelry.

Cat's-eye chrysoberyl: Chrysoberyl is the gem that exhibits the finest "cat's-eye." When the name "cat's-eye" is used alone, it is understood that the speaker is referring to a cat's-eye chrysoberyl. When the speaker is referring to any other gem with a cat's-eye, the name of that gem is used after the words "cat's-eye." For example: "cat's-eye apatite." This green cat's-eye chrysoberyl was produced in Sri Lanka and is about 5.6 x 4 millimeters in size.

Chatoyance occurs in stones that contain a large number of very thin parallel inclusions within the stone, known as a "silk." The light reflects from these inclusions to form a thin band across the surface of the stone. The band of light always occurs at right angles to the length of the parallel inclusions. These inclusions can be crystals, hollow tubes, or other linear structures that are present throughout the stone and are usually aligned with a crystallographic axis. Needle-like crystals of rutile and hematite are well-known for producing a cat's-eye in many specimens.

Tiger's-eye cabochons: Two "honey" tiger's-eye cabochons with very nice chatoyance. These stones are 8 x 6 millimeter ovals and cost a few dollars each. They were cut from material mined in the Central African Republic. Tiger's-eye is the only chatoyant stone that is produced in large quantities in calibrated sizes. That makes it the only chatoyant stone commonly seen in mass-production jewelry stores. It is very popular in men's rings and cufflinks.

Chatoyance - Eye of the Cat

The name "chatoyance" originates from the French word "*chatoyer*," which means "to shine like a cat's eye." The analogy also matches the way a cat's pupils will narrow to a thin slit under bright light.

Cat's eye with a slit pupil: This cat's eyes have responded to bright light by closing their pupils to a narrow slit. This is where the "cat's-eye" gemstone name originates.

Cat's-eye actinolite: Two cat's-eye actinolite cabochons cut from material mined in Tanzania. Actinolite is an amphibole mineral that often has a fibrous internal texture that serves as a "silk" and produces very strong eyes. These stones are 7-millimeter rounds.

Understanding Chatoyance

A good analogy for the chatoyance effect is how light reflects to form a line across the surface of a spool of silk thread. Viewing and moving a spool of silk thread under a beam of incident light can be a useful way to develop an understanding of how parallel inclusions produce an "eye" within a gemstone.

Cat's-eye on silk thread: These spools of silk thread exhibit chatoyance. Light shining on the spools reflects from the lustrous silk threads. As a result, a line of light similar to the "eye" of a gemstone forms across the portion of the spool where the tangent of the spool's surface is at a right angle to the observer's line of sight.

In a chatoyant gemstone, the band of light will move back and forth beneath the surface of the gem as it is turned under a beam of incident light. The band will also move if the position of the light is moved, or the observer moves his head to view the stone from a different angle. The motion of the cat's-eye across the top of the gem is one of the things that makes these stones so interesting, beautiful, and desired by many people.

Cat's-eye apatite: Apatite is a mineral that often produces a very nice cat's-eye on a gemmy base color. The golden-green cabochon on the left was cut from material mined in Kenya. It is a 7 x 6 millimeter semitransparent oval with a nice eye and a great base color. The golden-brown stone on the right was also cut from material mined in Kenya. It is a 9 x 7 millimeter translucent oval.

Many Types of Gems Exhibit Chatoyance

Many types of gems exhibit chatoyance; however, it is not present in every specimen of those gems and is usually only seen in a minority of specimens. Gems with well-known chatoyance include the following:chrysoberyl, tiger's-eye, actinolite, apatite, beryl (aquamarine, heliodor, emerald), beryllonite, cerussite, danburite, diaspore, diopside, enstatite, garnet, iolite, kyanite, moonstone, peridot, pezzottaite, prehnite, quartz, rutile, sillimanite, scapolite, spinel, topaz, tourmaline, and zircon.

Cat's-eye sillimanite: This cabochon is an 11 x 8 millimeter oval cut from a purplish brown sillimanite mined in Madagascar. It has a distinct pink eye on a background that is nearly opaque.

Cat's-eye aquamarine: Rare translucent aquamarines are known to produce cat's-eyes, although the eyes are generally not very strong. This greenish blue stone is about 11 x 8 millimeters in size and was cut from material mined in Madagascar.

Milk and Honey

Some gems with a highly developed chatoyance can appear to be made of two different materials when illuminated from the proper direction with respect to the observer's eye. In these stones, the cat's-eye will appear to divide the stone into a zone of light-colored material on one side of the eye and dark-colored material on the other. This phenomenon is known as the "milk-and-honey" effect. A photo of cat's-eye chrysoberyl showing the milk-and-honey effect is included below.

Milk-and-honey chrysoberyl: Chrysoberyl is the gem that exhibits the finest cat's-eye. When the term "cat's-eye" is used without a mineral name as a modifier, the speaker is most likely referring to chrysoberyl. This specimen exhibits the "milk-and-honey" effect - when properly oriented, the stone has two apparently different colors on each side of the cat's-eye line. This green cat's-eye chrysoberyl was produced in Sri Lanka and is about 5.6 x 4 millimeters in size. The stone in this photo is the same stone shown in the photo at the top of the page.

Cat's-eye heliodor: This golden-yellow heliodor beryl was mined in Madagascar and cut into a 10 x 8 millimeter oval. It has a beautiful translucent color and a faint eye.

Cutting the Cat's-Eye

Anyone who cuts cabochons with the intent of producing a cat's-eye stone must first find a material with a silk of inclusions that is capable of producing an eye. Then the cutter must examine the rough and orient the stone so that the silk will be parallel to the bottom of the stone within the finished cabochon, and the parallel inclusions will be parallel to the long axis of the cabochon. If this proper orientation is not achieved, the stone will have an off-center eye or no eye at all.

Cat's-eye tourmaline: Tourmaline often contains a silk of coarse tubes that can produce a very strong cat's-eye. In these stones you can see how the silk crosses the stone at right angles to the cat's-eye. A skilled cutter must assess the stone before cutting so that this orientation can be achieved. These colorful cabochons were cut from material mined in Brazil. The pink stone is an 8.5 millimeter round, and the green stone is a 10 x 7 millimeter oval.

Assessing the Quality of a Cat's-Eye Gem

The best cat's-eye gems have an eye that meets the following criteria:

- It is clearly visible
- It symmetrically bisects the cabochon

- It contrasts sharply with the stone's body color
- It moves smoothly as the stone is turned under light

If a stone features all of the above criteria and has an exceptional body color, then you have a fantastic cat's-eye gem. The importance of a desirable body color cannot be overstated.

Cat's-eye scapolite: This scapolite cabochon, cut from material mined in India, is especially interesting. It contains a very coarse silk that not only produces a cat's-eye, but also serves as a diffraction grating that produces a strong display of spectral colors. The stone is a 12 x 9 millimeter oval.

Diffraction in Cat's-Eye

Rare chatoyant specimens will have a coarse silk with just the right spacing to serve as a diffraction grating. These specimens will produce not only a cat's-eye, but also a display of spectral colors caused by light passing through the coarse silk and being diffracted into the colors of the spectrum. More familiar examples of diffraction are rainbows produced by sunlight diffracted by raindrops, or opal's play-of-color caused by light diffracted by tiny spheres of opal within the gem. The specimen of cat's-eye scapolite in the photo on this page is an example. (Another stone that exhibits diffraction is iris agate).

References

- Martin Meredith (2007). Diamonds, Gold, and War: The British, the Boers, and the Making of South Africa. New York: Public affairs. p. 16. ISBN 1-58648-473-7.
- Gemology: investopedia.com, Retrieved 28 March 2018
- Walker, J. (1979). "Optical absorption and luminescence in diamond". Rep. Prog. Phys. 42(10): 1605. doi:10.1088/0034-4885/42/10/001.
- What-are-gemstones: eternaltools.com, Retrieved 18 May 2018
- "A Contribution to the Understanding of Blue Fluorescence on the Appearance of Diamonds" (PDF). GIA. GIA. Winter 1997. Retrieved March 1, 2007.
- The-five-cardinal-gemstones: worldatlas.com, Retrieved 28 April 2018
- Nassau, K.; Broyer, A. M. (1962). "Application of Czochralski Crystal-Pulling Technique to High-Melting Oxides". Journal of the American Ceramic Society. 45 (10): 474. doi:10.1111/j.1151-2916.1962.tb11037.x.
- Diamond-gemstone: britannica.com, Retrieved 26 June 2018

- John Hays Hammond (1974). The Autobiography of John Hays Hammond. Ayer Publishing. p. 205. ISBN 0-405-05913-2.
- Asterism-in-gems: jewelinfo4u.com, Retrieved 15 July 2018
- Harman, Alang Kasim; Ninomiya, Susumu; Adachi, Sadao (1994). "Optical constants of sapphire (alpha-Al2O3) single crystals". Journal of Applied Physics. 76 (12): 8032–8036. Bibcode:1994JAP....76.8032H. doi:10.1063/1.357922.

Chapter 6

Crystallography: A Comprehensive Study

Crystallography is the science concerned with the determination of the arrangement of atoms in a crystalline solid. In order to study the role of crystallography, it is necessary to understand the fundamentals of lattice constant, crystal systems and their types, crystal family and crystal habit, besides others which have been extensively dealt in this chapter.

The study of the growth, shape, and geometric character of these beautiful forms of minerals is called crystallography.

Crystallography began with the study of crystals, like quartz. Today, crystallographers study the atomic architecture of any material that can form an orderly solid - from diamonds to viruses. They also investigate a wide variety of other materials, such as amorphous thin films, membranes, liquid crystals, fibers, glasses, liquids, gases and quasicrystals.

Because many crystallographers use x-rays to study crystals, the field is often called "x-ray crystallography." But modern crystallographers use many other methods as well. Atomic force microscopy, neutron diffraction, electron crystallography, molecular modeling, high- and low-temperature studies, high-pressure diffraction and micro-gravity experiments in space are all methods used by crystallographers to unlock the secrets of structure and function.

Three Types of Crystallography

The three types of radiation—X-rays, electrons, and neutrons—interact with the specimen in different ways. X-rays interact with the spatial distribution of the valence electrons, while electronsare charged particles and therefore feel the total charge distribution of both the atomic nuclei and the surrounding electrons. Neutrons are scattered by the atomic nuclei through the strong nuclear forces, but in addition, the magnetic moment of neutrons is non-zero. They are therefore also scattered by magnetic fields. Because of these different forms of interaction, the three types of radiation are suitable for different crystallographic studies.

Theory

An image of a small object is made using a lens to focus the beam, similar to a lens in a microscope. However, the wavelength of visible light (about 4000 to 7000 ångström) is three orders of magnitude longer than the length of typical atomic bonds and atoms themselves. Therefore, obtaining information about the spatial arrangement of atoms requires the use of radiation with shorter wavelengths, such as X-ray or neutron beams. Employing shorter wavelengths implied abandoning microscopy and true imaging, however, because there exists no material from which a lens capable of focusing this type of radiation can be created. Scientists have had some success focusing X-rays with microscopic Fresnel zone plates made from gold, and by critical-angle reflection inside long tapered capillaries. Diffracted X-ray or neutron beams cannot be focused to produce images, so the sample structure must be reconstructed from the diffraction pattern. Sharp features in the diffraction pattern arise from periodic, repeating structure in the sample, which are often very strong due to coherent reflection of many photons from many regularly spaced instances of similar structure, while non-periodic components of the structure result in diffuse (and usually weak) diffraction features - areas with a higher density and repetition of atom order tend to reflect more light toward one point in space when compared to those areas with fewer atoms and less repetition.

Because of their highly ordered and repetitive structure, crystals give diffraction patterns of sharp Bragg reflectionspots, and are ideal for analyzing the structure of solids.

Notation

- Coordinates in *square brackets* such as [100] denote a direction vector (in real space).
- Coordinates in *angle brackets* or *chevrons* such as <100> denote a family of directions which are related by symmetry operations. In the cubic crystal systemfor example, <100> would mean [100], [010], [001] or the negative of any of those directions.
- Miller indices in parentheses such as (100) denote a plane of the crystal structure, and regular repetitions of that plane with a particular spacing. In the cubic system, the normal to the (hkl) plane is the direction [hkl], but in lower-symmetry cases, the normal to (hkl) is not parallel to [hkl].
- Indices in curly brackets or braces such as {100} denote a family of planes and their normals which are equivalent in cubic materials due to symmetry operations, much the way angle brackets denote a family of directions. In non-cubic materials, <hkl> is not necessarily perpendicular to {hkl}.

Techniques

Some materials that have been analyzed crystallographically, such as proteins, do not occur naturally as crystals. Typically, such molecules are placed in solution and allowed to slowly crystallize through vapor diffusion. A drop of solution containing the molecule, buffer, and precipitants is sealed in a container with a reservoir containing a hygroscopic solution. Water in the drop diffuses to the reservoir, slowly increasing the concentration and allowing a crystal to form. If the concentration were to rise more quickly, the molecule would simply precipitate out of solution, resulting in disorderly granules rather than an orderly and hence usable crystal.

Once a crystal is obtained, data can be collected using a beam of radiation. Although many universities that engage in crystallographic research have their own X-ray producing equipment, synchrotrons are often used as X-ray sources, because of the purer and more complete patterns such sources can generate. Synchrotron sources also have a much higher intensity of X-ray beams, so data collection takes a fraction of the time normally necessary at weaker sources. Complementary neutron crystallography techniques are used to identify the positions of hydrogen atoms, since X-rays only interact very weakly with light elements such as hydrogen.

Producing an image from a diffraction pattern requires sophisticated mathematics and often an iterative process of modelling and refinement. In this process, the mathematically predicted diffraction patterns of an hypothesized or "model" structure are compared to the actual pattern generated by the crystalline sample. Ideally, researchers make several initial guesses, which through refinement all converge on the same answer. Models are refined until their predicted patterns match to as great a degree as can be achieved without radical revision of the model. This is a painstaking process, made much easier today by computers.

The mathematical methods for the analysis of diffraction data only apply to *patterns,* which in turn result only when waves diffract from orderly arrays. Hence crystallography applies for the most part only to crystals, or to molecules which can be coaxed to crystallize for the sake of measurement. In spite of this, a certain amount of molecular information can be deduced from patterns that are generated by fibers and powders, which while not as perfect as a solid crystal, may exhibit a degree of order. This level of order can be sufficient to deduce the structure of simple molecules, or to determine the coarse features of more complicated molecules. For example, the double-helical structure of DNA was deduced from an X-ray diffraction pattern that had been generated by a fibrous sample.

In Materials Science

Crystallography is used by materials scientists to characterize different materials. In single crystals, the effects of the crystalline arrangement of atoms are often easy to see macroscopically, because the natural shapes of crystals reflect the atomic structure. In addition, physical properties are often controlled by crystalline defects. The understanding of crystal structures is an important prerequisite for understanding crystallographic defects. Mostly, materials do not occur as a single crystal, but in poly-crystalline form (i.e., as an aggregate of small crystals with different orientations). Because of this, the powder diffraction method, which takes diffraction patterns of polycrystalline samples with a large number of crystals, plays an important role in structural determination.

Other physical properties are also linked to crystallography. For example, the minerals in clay form small, flat, platelike structures. Clay can be easily deformed because the platelike particles can slip along each other in the plane of the plates, yet remain strongly connected in the direction perpendicular to the plates. Such mechanisms can be studied by crystallographic texture measurements.

In another example, iron transforms from a body-centered cubic (bcc) structure to a face-centered cubic (fcc) structure called austenite when it is heated. The fcc structure is a close-packed structure unlike the bcc structure; thus the volume of the iron decreases when this transformation occurs.

Crystallography is useful in phase identification. When manufacturing or using a material, it is generally desirable to know what compounds and what phases are present in the material, as their composition, structure and proportions will influence the material's properties. Each phase has a characteristic arrangement of atoms. X-ray or neutron diffraction can be used to identify which patterns are present in the material, and thus which compounds are present. Crystallography covers the enumeration of the symmetry patterns which can be formed by atoms in a crystal and for this reason is related to group theory and geometry.

Crystal

Crystal is any solid material in which the component atoms are arranged in a definite pattern and whose surface regularity reflects its internal symmetry.

Categories of Crystals

Crystals are classified in general categories, such as insulators, metals, semiconductors, and molecular solids. A single crystal of an insulator is usually transparent and resembles a piece of glass. Metals are shiny unless they have rusted. Semiconductors are sometimes shiny and sometimes transparent but are never rusty. Many crystals can be classified as a single type of solid, while others have intermediate behaviour. Cadmium sulfide (CdS) can be prepared in pure form and is an excellent insulator; when impurities are added to cadmium sulfide, it becomes an interesting semiconductor. Bismuth (Bi) appears to be a metal, but the number of electrons available for electrical conduction is similar to that of semiconductors. In fact, bismuth is called a semimetal. Molecular solids are usually crystals formed from molecules or polymers. They can be insulating, semiconducting, or metallic, depending on the type of molecules in the crystal. New molecules are continuously being synthesized, and many are made into crystals. The number of different crystals is enormous.

Structure

Crystals can be grown under moderate conditions from all 92 naturally occurring elements except helium, and helium can be crystallized at low temperatures by using 25 atmospheres of pressure. Binary crystals are composed of two elements. There are thousands of binary crystals; some examples are sodium chloride (NaCl), alumina (Al_2O_3), and ice (H_2O). Crystals can also be formed with three or more elements.

The Unit Cell

A basic concept in crystal structures is the unit cell. It is the smallest unit of volume that permits identical cells to be stacked together to fill all space. By repeating the pattern of the unit cell over and over in all directions, the entire crystal lattice can be constructed. A cube is the simplest example of a unit cell. Two other examples. The first is the unit cell for a face-centred cubic lattice, and the second is for a body-centred cubic lattice. These structures are explained in the following paragraphs. There are only a few different unit-cell shapes, so many different crystals share a single unit-cell type. An important characteristic of a unit cell is the number of atoms it contains. The total number of atoms in the entire crystal is the number in each cell multiplied by the number of unit cells. Copper and aluminum (Al) each have one atom per unit cell, while zinc (Zn) and sodium chloride have two. Most crystals have only a few atoms per unit cell, but there are some exceptions. Crystals of polymers, for example, have thousands of atoms in each unit cell.

Structures of Metals

The elements are found in a variety of crystal packing arrangements. The most common lattice structures for metals are those obtained by stacking the atomic spheres into the most compact arrangement. There are two such possible periodic arrangements. In each, the first layer has the atoms packed into a plane-triangular lattice in which every atom has six immediate neighbours. Figure below shows this arrangement for the atoms labeled *A*. The second layer is shaded in the figure. It has the same plane-triangular structure; the atoms sit in the holes formed by the first layer. The first layer has two equivalent sets of holes, but the atoms of the second layer can occupy only one set. The third layer, labeled *C*, has the same structure, but there are two choices for selecting the holes that the atoms will occupy. The third layer can be placed over the atoms of the first layer, generating an alternate layer sequence ABABAB..., which is called the hexagonal- closest-packed (hcp) structure. Cadmium and zinc crystallize with this structure. The second possibility is to place the atoms of the third layer over those of neither of the first two but instead over the set of holes in the first layer that remains unoccupied. The fourth layer is placed over the first, and so there is a three-layer repetition ABCABCABC..., which is called the face-centred cubic (fcc), or cubic-closest-packed, lattice. Copper, silver (Ag), and gold (Au) crystallize in fcc lattices. In the hcp and the fcc structures the spheres fill 74 percent of the volume, which represents the closest possible packing of spheres. Each atom has 12 neighbours. The number of atoms in a unit cell is two for hcp structures and one for fcc. There are 32 metals that have the hcp lattice and 26 with the fcc. Another possible arrangement is the body-centred cubic (bcc) lattice, in which each atom has eight neighbours arranged at the corners of a cube. Figure A below shows the cesium chloride (CsCl) structure, which is a cubic arrangement. If all atoms in this structure are of the same species, it is a bcc lattice. The spheres occupy 68 percent of the volume. There are 23 metals with the bcc arrangement. The sum of these three numbers (32 + 26 + 23) exceeds the number of elements that form metals (63), since some elements are found in two or three of these structures.

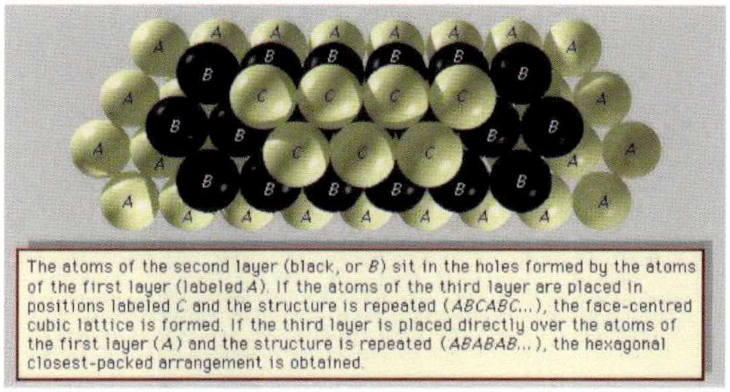

Figure: Stacking of spheres in closest-packed arrangements.

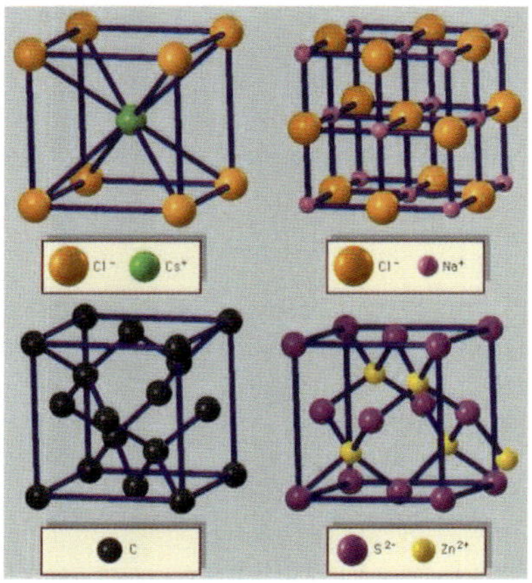

Figure: *Crystal structures*. There is an equal number of the two types of ions in the unit cell of the (A) cesium chloride, (B) sodium chloride, and (D) zinc blende arrangements. The diamond arrangement is shown in (C). If both atoms are identical in (A), the structure is body-centred cubic.

The fcc structure is also found for crystals of the rare gas solids neon (Ne), argon (Ar), krypton (Kr), and xenon (Xe). Their melting temperatures at atmospheric pressure are: Ne, 24.6 K; Ar, 83.8 K; Kr, 115.8 K; and Xe, 161.4 K.

Structures of Nonmetallic Elements

The elements in the fourth row of the periodic table—carbon, silicon, germanium (Ge), and α-tin (α-Sn)—prefer covalent bonding. Carbon has several possible crystal structures. Each atom in the covalent bond has four first-neighbours, which are at the corners of a tetrahedron. This arrangement is called the diamond lattice and is shown in figure C above. There are two atoms in a unit cell, which is fcc. Large crystals of diamond are valuable gemstones. The crystal has other interesting properties; it has the highest sound velocity of any solid and is the best conductor of heat. Besides diamond, the other common form of carbon is graphite, which is a layered material. Each carbon atom has three coplanar near neighbours, forming an arrangement called the honeycomb lattice. Three-dimensional graphite crystals are obtained by stacking similar layers.

Another form of crystalline carbon is based on a molecule with 60 carbon atoms called buckminsterfullerene (C_{60}). The molecular shape is spherical. Each carbon is bonded to three neighbours, as in graphite, and the spherical shape is achieved by a mixture of 12 rings with five sides and 20 rings with six sides. Similar structures were first visualized by the American architect R. Buckminster Fuller for geodesic domes. The C_{60} molecules, also called buckyballs, are quite strong and almost incompressible. Crystals are formed such that the balls are arranged in an fcc lattice with a one-nanometre spacing between the centres of adjacent balls. The similar C_{70} molecule has the shape of a rugby ball; C_{70}molecules also form an fcc crystal when stacked together. The solid fullerenes form molecular crystals, with weak binding—provided by van der Waals interactions—between the molecules.

Many elements form diatomic gases: hydrogen (H), oxygen (O), nitrogen (N), fluorine (F), chlorine (Cl), bromine (Br), and iodine (I). When cooled to low temperature, they form solids of diatomic molecules. Nitrogen has the hcp structure, while oxygen has a more complex structure.

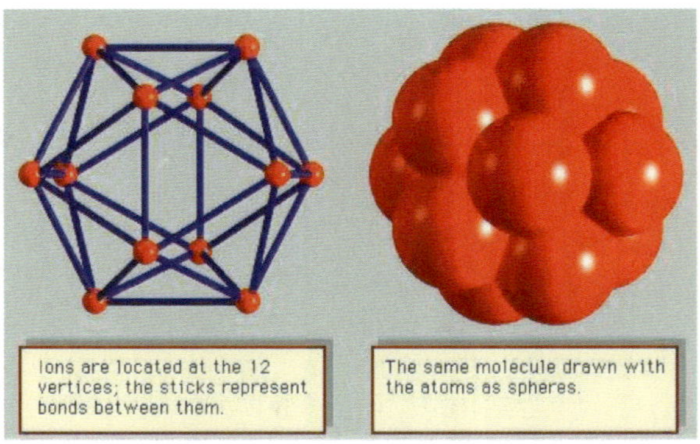

Figure: The icosahedral arrangement of boron molecules.

The most interesting crystal structures are those of elements that are neither metallic, covalent, nor diatomic. Although boron (B) and sulfur (S) have several different crystal structures, each has one arrangement in which it is usually found. Twelve boron atoms form a molecule in the shape of an icosahedron. Crystals are formed by stacking the molecules. The β-rhombohedral structure of boron has seven of these icosahedral molecules in each unit cell, giving a total of 84 atoms. Molecules of sulfur are usually arranged in rings; the most common ring has eight atoms. The typical structure is α-sulfur, which has 16 molecules per unit cell, or 128 atoms. In the common crystals of selenium (Se) and tellurium (Te), the atoms are arranged in helical chains, which stack like cordwood. However, selenium also makes eight-atom rings, similar to sulfur, and forms crystals from them. Sulfur also makes helical chains, similar to selenium, and stacks them together into crystals.

Structures of Binary Crystals

Binary crystals are found in many structures. Some pairs of elements form more than one structure. At room temperature, cadmium sulfide may crystallize either in the zinc blende or wurtzite structure. Alumina also has two possible structures at room temperature, α-alumina (corundum) and β-alumina. Other binary crystals exhibit different structures at different temperatures. Among the most complex crystals are those of silicon dioxide (SiO_2), which has seven different structures at various temperatures and pressures; the most common of these structures is quartz. Some pairs of elements form several different crystals in which the ions have different chemical valences. Cadmium (Cd) and phosphorus (P) form the crystals Cd_3P_2, CdP_2, CdP_4, Cd_7P_{10}, and Cd_6P_7. Only in the first case are the ions assigned the expected chemical valences of Cd^{2+} and P^{3-}.

Among the binary crystals, the easiest structures to visualize are those with equal numbers of the two types of atoms. The structure of sodium chloride is based on a cube. To construct the lattice, the sodium and chlorine atoms are placed on alternate corners of a cube, and the structure is repeated. The structure of the sodium atoms alone, or the chlorine atoms alone, is fcc and defines the unit cell. The sodium chloride structure thus is made up of two interpenetrating fcc lattices. The cesium chloride lattice (Figure A) is based on the bcc structure; every other atom is cesium or chlorine. In this case, the unit cell is a cube. The third important structure for *AB* (binary) lattices is zinc blende (Figure D). It is based on the diamond structure, where every other atom is *A* or *B*. Many binary semiconductors have this structure, including those with one atom from the third (boron, aluminum, gallium [Ga], or indium [In]) and one from the fifth (nitrogen, phosphorus, arsenic [As], or antimony [Sb]) column of the periodic table (GaAs, InP, etc.). Most of the chalcogenides (O, S, Se, Te) of cadmium and zinc (CdTe, ZnSe, ZnTe, etc.) also have the zinc blende structure. The mineral zinc blende is ZnS; its unit cell is also fcc. The wurtzite structure is based on the hcp lattice, where every other atom is *A* or *B*. These four structures comprise most of the binary crystals with equal numbers of cations and anions.

The fullerene molecule forms binary crystals M_xC_{60} with alkali atoms, where M is potassium (K), rubidium (Rb), or cesium (Cs). The fullerene molecules retain their spherical shape, and the alkali atoms sit between them. The subscript x can take on several values. A compound with $x = 6$ (*e.g.*, K_6C_{60}) is an insulator with the fullerenes in a bcc structure. The case $x = 4$ is an insulator with the body-centred tetragonal structure, while the case $x = 3$ is a metal with the fullerenes in an fcc structure. K_3C_{60}, Rb_3C_{60}, and Cs_3C_{60} are superconductors at low temperatures.

Alloys

Alloys are solid mixtures of atoms with metallic properties. The definition includes both amorphous and crystalline solids. Although many pairs of elements will mix together as solids, many pairs will not. Almost all chemical entities can be mixed in liquid form. But cooling a liquid to form a solid often results in phase separation; a polycrystalline material is obtained in which each grain is purely one atom or the other. Extremely rapid cooling can produce an amorphous alloy. Some pairs of elements form alloys that are metallic crystals. They have useful properties that differ from those exhibited by the pure elements. For example, alloying makes a metal stronger; for this reason alloys of gold, rather than the pure metal, are used in jewelry.

Atoms tend to form crystalline alloys when they are of similar size. The sizes of atoms are not easy to define, however, because atoms are not rigid objects with sharp boundaries. The outer part of an atom is composed of electrons in bound orbits; the average number of electrons decreases gradually with increasing distance from the nucleus. There is no point that can be assigned as the precise radius of the atom. Scientists have discovered, however, that each atom in a solid has a characteristic radius that determines its preferred separation from neighbouring atoms. For most types of atom this radius is constant, even in different solids. An empirical radius is assigned to each atom for bonding considerations, which leads to the concept of atomic size. Atoms readily make crystalline alloys when the radii of the two types of atoms agree to within roughly 15 percent.

Two kinds of ordering are found in crystalline alloys. Most alloys at low temperature are binary crystals with perfect ordering. An example is the alloy of copper and zinc. Copper is fcc, whereas zinc is hcp. A 50-percent-zinc–50-percent-copper alloy has a different structure—β-brass. At low temperatures it has the cesium chloride structure: a bcc lattice with alternating atoms of copper and zinc and a cubic unit cell. If the temperature is raised above 470°C, however, a phase transition to another crystalline state occurs. The ordering at high temperature is also bcc, but now each site has equal probability of having a copper or zinc atom. The two types of atoms randomly occupy each site, but there is still long-range order. At all temperatures, thousands of atoms away from a site, the location of the atom site can be predicted with certainty. At temperatures below 470°C one also knows whether that site will be occupied by a copper or zinc atom, while above 470°C there is an equal likelihood of finding either atom. The high-temperature phase is crystalline but disordered. The disorder phase is obtained through a partial melting, not into a liquid state but into a less ordered one. This behaviour is typical of metal alloys. Other common alloys are steel, an alloy of iron and carbon; stainless steel, an alloy of iron, nickel (Ni), and chromium (Cr); pewter and solder, alloys of tin and lead (Pb); and britannia metal, an alloy of tin, antimony, and copper.

Crystal Defects

A crystal is never perfect; a variety of imperfections can mar the ordering. A defect is a small imperfection affecting a few atoms. The simplest type of defect is a missing atom and is called a vacancy. Since all atoms occupy space, extra atoms cannot be located at the lattice sites of other atoms, but they can be found between them; such atoms are called interstitials. Thermal vibrations may cause an atom to leave its original crystal site and move into a nearby interstitial site, creating a vacancy-interstitial pair. Vacancies and interstitials are the types of defects found in a pure crystal. In another defect, called an impurity, an atom is present that is different from the host crystal atoms. Impurities may either occupy interstitial spaces or substitute for a host atom in its lattice site.

There is no sharp distinction between an alloy and a crystal with many impurities. An alloy results when a sufficient number of impurities are added that are soluble in the host metal. However, most elements are not soluble in most crystals. Crystals generally can tolerate a few impurities per million host atoms. If too many impurities of the insoluble variety are added, they coalesce to form their own small crystallite. These inclusions are called precipitates and constitute a large defect.

Germanium is a common impurity in silicon. It prefers the same tetrahedral bonding as silicon and readily substitutes for silicon atoms. Similarly, silicon is a common impurity in germanium. No large crystal can be made without impurities; the purest large crystal ever grown was made of germanium. It had about 10^{10} impurities in each cubic centimetre of material, which is less than one impurity for each trillion atoms.

Impurities often make crystals more useful. In the absence of impurities, α-alumina is colorless. Iron and titanium impurities impart to it a blue color, and the resulting gem-quality mineral is known as sapphire. Chromium impurities are responsible for the red color characteristic of rubies, the other gem of α-alumina. Pure semiconductors rarely conduct electricity well at room temperatures. Their ability to conduct electricity is caused by impurities. Such impurities are deliberately added to silicon in the manufacture of integrated circuits. In fluorescent lamps the visible light is emitted by impurities in the phosphors (luminescent materials).

Other imperfections in crystals involve many atoms. Twinning is a special type of grain boundary defect, in which a crystal is joined to its mirror image. Another kind of imperfection is a dislocation, which is a line defect that may run the length of the crystal. One of the many types of dislocations is due to an extra plane of atoms that is inserted somewhere in the crystal structure. Another type, called an edge dislocation, is shown in Figure. This line defect occurs when there is a missing row of atoms. In the figure the crystal arrangement is perfect on the top and on the bottom. The defect is the row of atoms missing from region *b*. This mistake runs in a line that is perpendicular to the page and places a strain on region *a*.

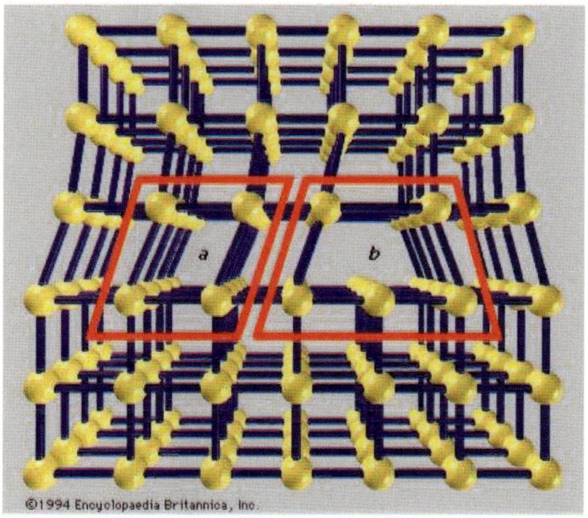

Figure: *Crystalline lattice defect*. An edge dislocation occurs when there is a missing row of atoms as shown in region *b*.Region *a* is strained.

Dislocations are formed when a crystal is grown, and great care must be taken to produce a crystal free of them. Dislocations are stable and will exist for years. They relieve mechanical stress. If one presses on a crystal, it will accommodate the induced stress by growing dislocations at the surface, which gradually move inward. Dislocations make a crystal mechanically harder. When a metal bar is cold-worked by rolling or hammering, dislocations and grain boundaries are introduced; this causes the hardening.

Determination of Crystal Structures

Crystal structures are determined by scattering experiments using a portion of the crystal as the target. A beam of particles is sent toward the target, and upon impact some of the particles scatter from the crystal and ricochet in various directions. A measurement of the scattered particles provides raw data, which is then computer-processed to give a picture of the atomic arrangements. The positions are then inferred from the computer-analyzed data.

Max von Laue first suggested in 1912 that this measurement could be done using X rays, which are electromagnetic radiation of very high frequency. High frequencies are needed because these waves have a short wavelength. Von Laue realized that atoms have a spacing of only a few angstroms (1 angstrom [Å] is 10^{-10} metre, or 3.94×10^{-9} inch). In order to measure atomic arrangements, the particles scattering from the target must also have a wavelength of a few angstroms. X rays are required when the beam consists of electromagnetic radiation. The X rays only scatter in certain directions, and there are many X rays associated with each direction. The scattered particles appear in spots corresponding to locations where the scattering from each identical atom produces an outgoing wave that has all the wavelengths in phase. Figure shows incoming waves in phase. The scattering from atom A_2 has a longer path than that from atom A_1. If this additional path has a length ($AB + BC$) that is an exact multiple of the wavelength, then the two outgoing waves are in phase and reinforce each other. If the scattering angle is changed slightly, the waves no longer add coherently and begin to cancel one another. Combining the scattered radiation from all the atoms in the crystal causes all the outgoing waves to add coherently in certain directions and produce a strong signal in the scattered wave. If the extra path length ($AB + BC$) is five wavelengths, for example, the spot appears in one place. If it is six wavelengths, the spot is elsewhere. Thus, the different spots correspond to the different multiples of the wavelength of the X ray. The measurement produces two types of information: the directions of the spots and their intensity. This information is insufficient to deduce the exact crystal structure, however, as there is no algorithm by which the computer can go directly from the data to the structure. The crystallographer must propose various structures and compute how they would scatter the X rays. The theoretical results are compared with the measured one, and the theoretical arrangement is chosen that best fits the data. Although this procedure is fast when there are only a few atoms in a unit cell, it may take months or years for complex structures. Some protein molecules, for instance, have hundreds of atoms. Crystals of the proteins are grown, and X rays are used to measure the structure. The goal is to determine how the atoms are arranged in the protein, rather than how the proteins are arranged in the crystal.

Beams of neutrons may also be used to measure crystal structure. The beam of neutrons is obtained by drilling a hole in the side of a nuclear reactor. The energetic neutrons created in nuclear fission escape through the hole. The motion of elementary particles is governed by quantum, or wave, mechanics. Each neutron has a wavelength that depends on its momentum. The scattering directions are determined by the wavelength, as is the case with X rays. The wavelengths for neutrons from a reactor are suitable for measuring crystal structures.

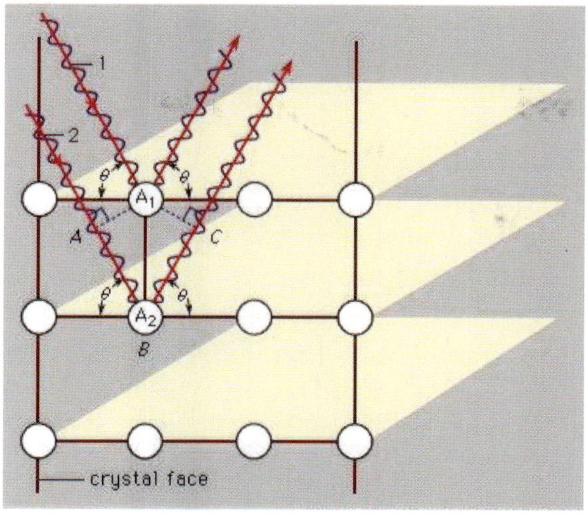

Figure: Incident rays (1 and 2) at angle θ on the planes of atoms in a crystal. Rays reinforce if their difference in path length ($AB + BC$) is an integer times the wavelength of the X ray.

X rays and neutrons provide the basis for two competing technologies in crystallography. Although they are similar in principle, the two methods have some differences. X rays scatter from the electrons in the atoms so that more electrons result in more scattering. X rays easily detect atoms of high atomic number, which have many electrons, but cannot readily locate atoms with few electrons. In hydrogen-bonded crystals, X rays do not detect the protons at all. Neutrons, on the other hand, scatter from the atomic nucleus. They scatter readily from protons and are excellent for determining the structure of hydrogen-bonded solids. One drawback to this method is that some nuclei absorb neutrons completely, and there is little scattering from these targets.

Beams of electrons can also be used to measure crystal structure, because energetic electrons have a wavelength that is suitable for such measurements. The problem with electrons is that they scatter strongly from atoms. Proper interpretation of the experimental results requires that an electron scatter only from one atom and leave the crystal without scattering again. Low-energy electrons scatter many times, and the interpretation must reflect this. Low-energy electron diffraction (LEED) is a technique in which a beam of electrons is directed toward the surface. The scattered electrons that reflect backward from the surface are measured. They scatter many times before leaving backward but mainly leave in a few directions that appear as “spots” in the measurements. An analysis of the varied spots gives information on the crystalline arrangement. Because the electrons are scattered strongly by the atoms in the first few layers of the surface, the measurement gives only the arrangements of atoms in these layers. It is assumed that the same structure is repeated throughout the crystal. Another scattering experiment involves electrons of extremely high energy. The scattering rate decreases as the energy of the electron increases, so that very energetic electrons usually scatter only once. Various electron microscopes are constructed on this principle.

Cleavage (Crystal)

Cleavage is the tendency of a crystal to break cleanly along distinct planes.

First cleavage is reproducible, meaning that a crystal can be broken along the same parallel plane over and over again. All cleavage must parallel a possible crystal face. This means that the crystal could have a crystal face parallel to its cleavage, but these faces are not always formed. All cleavage planes of a mineral must match that mineral's symmetry. And finally, the same mineral will always, always have the same cleavage.

The tendency for minerals to cleave or not and in which directions is very characteristic and therefore important to the identification of minerals. Cleavage is described in terms of how easy the cleavage is produced. From easiest to hardest to produce the terms are: perfect, imperfect, good, distinct, indistinct, and poor. Cleavage is said to be basal when it occurs perpendicular to the major axis of the mineral and prismatic when it occurs parallel to the major axis. Multiple cleavages that produce geometric polygons are referred to using the name of the geometric polygon, such as octahedral cleavage in the mineral fluorite, cubic cleavage in the mineral halite or rhombohedral cleavage in calcite (pictured). Cleavage occurs in minerals that have specific planes of weakness. These planes or directions are inherent in the structure of the mineral and form from a variety of factors.

Cleavage, being related to structure, is important at times in the correct identification of a mineral's symmetry.Remember, cleavage must obey the symmetry of the mineral and must be parallel to a possible crystal face. A mineral of the isometric symmetry class can either have no cleavage or at least three directions of identical cleavage that form a closed three dimensional polygon. A mineral of a uniaxial class (trigonal, tetragonal or hexagonal) will potentially have a cleavage perpendicular to the dominant axis and prismatic cleavage of either 3, 4 or 6 directions respectively, running parallel to the axis. Other cleavage directions are possible, but will always be controlled by the symmetry of the crystal. A biaxial mineral, those belonging to orthorhombic, monoclinic or triclinic classes, can not have more than two identical cleavage directions.

The angle between cleavages is also important to note and may be diagnostic. The pyroxene and amphibole groups of minerals are distinguished primarily by cleavage angle with the pyroxenes having a more acute angle. The angle may also help identify the type of cleavage. Three identical directions of cleavage in one mineral can only be either cubic cleavage, rhombic cleavage or prismatic (forming six sided prisms). If the angle between cleavage faces is 90 degrees, then the cleavage is cubic. If the angle is 60 degrees, then the cleavage is prismatic. Also, if the angle is something else and there are *three identical cleavages*, then the cleavage is rhombic.

The phyllosilicates are a group of minerals whose structure is based upon stacked layers. A natural cleavage plane is produced between these layers. Other minerals may have cleavage planes that are more related to bond strength. Weak bonds that all lie in a plane will produce a cleavage direction.

To identify cleavage in a mineral remember that it is always parallel to a possible crystal face, it is reproducible over and over again and that it may be seen as internal reflection planes. Cleavage can be observed without the specimen being cleaved all the way through as pictured above. Minerals with perfect cleavage will sometimes have a stairstep look around a broken section. Twinning may break a minerals cleavage at the twin plane and this should be kept in mind. Knowledge of a mineral's cleavage can be important in determining if a given specimen has been broken or not (a key characteristic in a mineral specimen's value). The related property of parting is thought by many to just be an example of poor cleavage. Many minerals lack any cleavage at all and will only show fractures.

Crystal Habit

Acicular crystals have a needle-like shape that tapers to a point or a blunt termination. Many acicular crystals can be clustered to produce fan-shaped or radially-shaped aggregates. The name acicular should be used when the length of an individual crystal is much greater than its width or diameter. Mineral examples include rutile, natrolite, millerite, and gypsum. This geode found in Kentucky contains thin needle-like crystals of millerite. This specimen is also a good example of the geodic habit.

Crystal habit is the tendency for specimens of a mineral to repeatedly grow into characteristic shapes. These shapes are influenced by the atomic structure of the mineral, but they can also be influenced by the environment of crystal growth. Regardless of influence, crystal habit shapes are characteristic of the mineral and displayed by many specimens of that mineral. Some examples:

- Botryoidal: malachite and hematite
- Banded: rhodochrosite and fluorite
- Striated: tourmaline and beryl
- Acicular: rutile, millerite, and tourmaline
- Dendritic: copper and pyrolusite
- Nodular: quartz, azurite, and hematite
- Prismatic: quartz, beryl, and tourmaline

Banded

Banded minerals have narrow layers or bands of different color and texture. These may be a response to changes in the composition of the growth liquid, the sedimentary process, or other conditions. Mineral examples: quartz(agate), malachite, rhodochrosite, and fluorite. The photo above shows rhodochrosite cabochons that display a banded habit. In one of the cabochons, the banded habit is actually an internal feature of a stalactitic habit. The cabochons were cut from material mined in Argentina, and the cabochon on the left is about two centimeters in length.

Crystal Habits and Forms	
Acicular	Hopper
Banded	Massive
Bladed	Nodular
Botryoidal	Octahedral
Columnar	Oolitic
Cubic	Pisolitic
Dendritic	Prismatic
Dodecahedral	Radiating
Drusy	Rosette
Fibrous	Stalactitic
Foliated	Striated
Geodic	Tabular
Granular	

Crystal habit is an external shape displayed by an individual crystal, but more often it is an external shape displayed by an aggregate of crystals. Crystal habit names are often adjectives that help convey the shape of a crystal or a group of crystals.

Bladed, cubic, fibrous, granular, prismatic, and radiating are names of crystal habits that quickly convey a generalized geometric appearance. Other crystal habit names are less familiar, but once a person learns the name it can easily be applied to mineral specimens to which that habit applies.

Many minerals have characteristic habits which can serve as clues to their identification. However, many minerals do not have a characteristic external shape, and the name of this habit is "massive". In addition, most specimens of any mineral do not display a characteristic habit.

In rare instances, one mineral might replace crystals of another mineral to assume its characteristic habit. Crystals formed in this situation are known as pseudomorphs.

Bladed

Bladed crystals are elongated. They are much longer than they are wide, and their width exceeds their depth. They are shaped like a straight sword or knife blade. Their ends sometimes taper to a point. They might exist as single crystals, a cluster of many parallel crystals, or radiating clusters of crystals. Mineral examples: kyanite, actinolite, and stibnite. These blue crystals of kyanite have a bladed habit. Kyanite crystals are interesting because they have a hardness of 4.5 to 5 parallel to the length of their blades, and a hardness of 6.5 to 7 across the width of their blades. This specimen is approximately seven centimeters across.

Crystal Form

"Crystal form" is a concept similar to "crystal habit". A crystal form is a solid crystalline object that is bounded by a set of flat faces that are related to one another by symmetry. Euhedral crystals are the best representations of a crystal form. Examples include:

- Cubic crystals of halite or fluorite
- Octahedral crystals of diamond or fluorite
- Dodecahedral crystals of garnet
- Hexagonal crystals of quartz or corundum

Columnar

Columnar crystals are long prisms with enough width that the name acicular (or needle-like) does not apply. A single "column" might contain multiple parallel crystals. Mineral examples: calcite, tourmaline, and gypsum. These enormous crystals of selenite gypsum have a columnar habit. They are in the "Cave of the Crystals" cavern, Chihuahua, Mexico (a person in the lower-right quadrant of the photo serves as scale). These are some of the largest well-formed crystals in the world.

Botryoidal

Botryoidal (also known as globular or mammillary) is derived from the Greek word "botruoeidēs", which means "bunch of grapes". This habit name is used for crystal aggregates that have a globular or rounded shape. Mineral examples: hematite, malachite, smithsonite, hemimorphite, variscite, quartz (chalcedony), and goethite. These green crystal aggregates of malachite have a botryoidal habit. This view spans an area of the specimen approximately five millimeters across.

Dendritic

Dendritic crystals form a branching pattern, much like the branches of a tree, the veins in a leaf, or the branching pattern of streams in a drainage basin. Mineral examples: copper, pyrolusite, and other manganese oxide minerals. These crystals of pyrolusite formed on a bedding surface of a piece of lithographic limestone collected near Solnhofen, Germany.

Cubic

Cubic crystals of pyrite. Fluorite and halite are two common minerals with a cubic shape. Cubes have six square faces and four-fold rotational symmetry around three axes. The photo shows cubic crystals of pyrite from Navajún, Rioja, Spain, that have grown in a marlstone. Specimen is approximately 4 inches (9.5 centimeters) across.

Drusy

Drusy is a habit name used for a surface that is covered with small crystals. The crystals themselves are referred to as a druse. Quartz is the most common mineral found as a druse. Other mineral examples: uvarovite garnet, malachite, and azurite. The rock in the photo above has a drusy surface because it is covered by a layer of uvarovite crystals. This rock was collected from the Saranovskii Mine, located in the Urals region of Russia. It is approximately 18 x 13 x 2 centimeters in size.

Dodecahedral

Dodecahedral garnet crystals about four millimeters across, from Idaho. A dodecahedron is any polyhedron with twelve flat faces. The dodecahedron is one of the most common forms for garnet crystals.

Fibrous

Fibrous is a habit name used when minerals occur in very fine fiber-like crystals. They are often so fine that they look like fine hair. The habit also includes aggregates made up of a large number of parallel or radial fibers. Mineral examples: actinolite, chrysotile, serpentine, and tremolite. The actinolite crystals on this rock have a fibrous habit. Because of their fibrous shape (a roughly 1:20 aspect ratio) and properties, fibrous crystals of actinolite are regulated as asbestos.

Geodic

Geodic is a habit in which mineral aggregates form a rounded or oblate mass by crystallization on the inside walls of a cavity. Concentric bands or layers of mineral crystals subsequently develop, gradually infilling the cavity without infilling it completely and with a crystal-lined central void. The specimen in the photo is a geode formed by the precipitation of banded agate to form the external wall and initial layers. The center of the geode is lined with quartzcrystals.

Foliated or Micaceous

Foliated (also known as Micaceous) is a sheet-like or layered structure. Minerals with a foliated habit are often able to be split into thin sheets. Members of the mica family are the best examples of a foliated habit. Clay minerals and graphite can be described as having this habit, but their foliation is on a microscopic scale. Mineral examples: muscovite, biotite, and chlorite. This specimen of muscovite exhibits a foliated habit. The mineral can easily be separated into very thin sheets. The specimen is approximately 5 centimeters across.

Hopper

Hopper crystals are partially formed crystals that have experienced more rapid growth on their outer edges than in the center of the crystal. This causes them to be well-developed on the outer edges but less developed or "hollow" in the center. Halite (shown above) is one of the best-known examples of a mineral that sometimes displays the hopper crystal habit. Other mineral examples: galena and ice.

Granular

Granular is the habit of a crystalline aggregate composed of many rounded or equant anhedral crystals of approximately the same size. The crystals might be loose with no interstitial material, or they might be interlocking such as calcite grains in a marble. Mineral examples: olivine, bornite, and scheelite. The photo shows a specimen of granular olivine in basalt. The olivine is in small rounded grains of about 2 to 4 millimeters in size with no interstitial material.

Nodular

Nodular is the name of a habit in which mineral crystals grow to form rounded or bulbous structures. The crystals are usually arranged in a radial structure within the nodule, even though the nodules may exhibit concentric banding. In the concentric banding, each layer is composed of crystals growing up and outward from the layer immediately below. Mineral examples: quartz (agate), azurite, hematite, realgar, and variscite. A nodule of variscite (bright green), crandallite (canary yellow), wardite (gray) and montgomeryite (dark green). The nodule is approximately eleven centimeters across and was found in Utah County, Utah.

Massive

Massive is the habit name used for masses of crystals that have no distinctive geometry. Most specimens of almost every mineral do not exhibit an obvious habit or obvious crystal form. Shown above is a specimen of lizardite serpentine from Wayne County, New York. The piece has no visible internal structure or characteristic external shape.

Oolitic

Oolitic minerals occur in crystalline aggregates that are rounded and less than about four millimeters in size. Oolites form by chemical precipitation from a solution. Similar to pisolitic, but oolites are much smaller than pisolites. Mineral examples: hematite and calcite. A specimen of oolitic iron ore approximately ten centimeters across. The largest oolites are a few millimeters across.

Octahedral

Octahedral diamond crystal, yellow in color and weighing 98.63 carats, recovered from the Jubilee (Yubileynaya) Pipe, Sakha Republic, Russia. The crystal measures approximately 29 x 28 x 27 millimeters and contains inclusions of olivine, graphite, and sulfide minerals. This diamond crystal is extremely interesting. Its surface is covered with growth lines and trigons. Octahedrons have eight triangular faces and three axes of four-fold rotational symmetry.

Prismatic

Prismatic is a habit name for minerals that form in elongated crystals with opposite faces normally parallel to one another. The crystals are often striated along their length (as in tourmaline) or across their width (as in quartz). Mineral examples: tourmaline, quartz, beryl, hornblende, augite, diopside, and topaz. Shown in the figure above are prismatic crystals of colorful tourmaline from Afghanistan with striations parallel to their long axis. The largest specimen in this photo is about three centimeters in length.

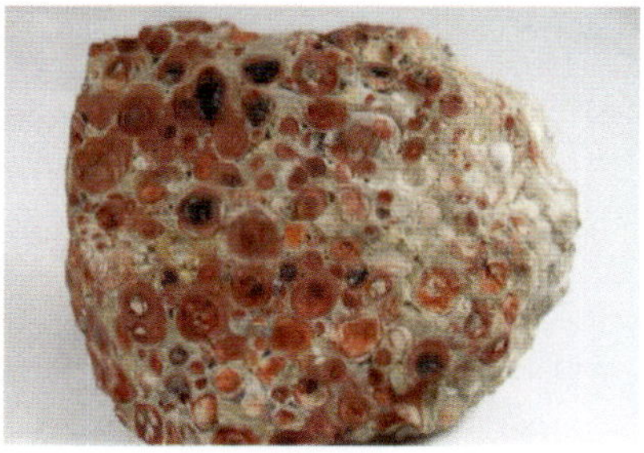

Pisolitic

Pisolitic minerals occur in crystalline aggregates that are rounded and about the size of peas. Individual pisolites are made up of many tiny radiating mineral crystals. They often develop a concentric structure formed when crystalline aggregate layers grow to enlarge the pisolites. Similar to oolitic, but pisolites are much larger than oolites. Mineral example: bauxite. The specimen of bauxite in the photo is about six centimeters across and contains numerous pisolites that are about the size of peas.

Rosettes

Rosettes are clusters of tabular crystals in a radial arrangement that have an external geometry that resembles a rose or flower. Barite and gypsumsometimes form crystals of this shape in sand to produce a rosette with a sandy appearance. Mineral examples: barite, gypsum, pyrite, and marcasite. The photo shows a "barite rose" which formed when a cluster of barite crystals grew in sand, incorporating many of the sand grains within each crystal.

Radiating

Radiating crystal aggregates grow outwards from a central point. They consist of multiple crystals growing in diverging directions. Mineral examples: wavellite, pyrite, rutile, and kyanite. The photo above shows a massive aggregate of kyanite crystals, many of which form radial clusters. The specimen is approximately five centimeters across.

Striations

Striations are fine, slightly indented lines that are present on the faces of some crystals. They always parallel a crystallographic axis and one of the edges of that crystal face. Mineral examples: pyrite, tourmaline, quartz, feldspar, euclase, and topaz. The photo shows a crystal of blue euclase with striations on its faces that parallel the long axis of the crystal. This specimen is also a good example of the prismatic crystal habit.

Stalactitic

Stalactitic is a habit name used for specimens that have formed as stalactites or stalagmites. The crystals often grow downwards or upwards in a cavity or cavern, yet they have a radial internal cross section. Mineral examples: calcite, malachite, goethite, and quartz. The photo shows a geode with stalactites of gem silica (inverted) from the Inspiration Mine, Gila County, Arizona.

Tabular

Tabular crystals are flat and plate-like. They have lengths and widths that are much larger than their thickness. An easy way to describe their shape is to compare them to a tablet computer or a tablet that you use to write notes. Mineral examples: feldspar, topaz, barite, and corundum. The photo shows tabular segments of a corundum crystal that separated along planes of parting.

Crystal System

There are six crystal systems. All minerals form crystals in one of these six systems. Although you may have seen more than six shapes of crystals, they're all variations of one of these six habits. Each system is defined by a combination of three factors:

- How many axes it has.
- The lengths of the axes.
- The angles at which the axes meet.

An axis is a direction between the sides. The shortest one is A. The longest is C. There is a B axis as well and sometimes a D axis.

Triclinic Crystal System

Triclinic Crystal Systems are those that have the inner shape of a trapezium (three inclined angles). The beauty of these crooked or bent crystals is that they have no right angles. The most striking example of where this is displayed is in Amazonite which forms in the shape of two trapeziums joined together. Other types are Kyanite, Labradorite, Rhodonite, Sunstone and Turquoise.

This crystal shape has a broad base which symbolizes a stable nature. However, it does need to be flipped upside down many times just to fit in. When it's upside down it's rather insecure looking. Both of these realities are present in a triclinic lifestyle as things seem to flip flop for you. Volatile, spontaneous and unpredictable, even to yourself, the course of your life is erratic. Fate continuously takes a hand in presenting you with unexpected new challenges.

Triclinic crystal system is also known as the rhombohedron system.

A rhombohedral crystal has six faces, 12 edges, and 8 vertices.

This system has the following forms:

- 3 Pinacoid
- 4 Pinacoid
- 10 Pinacoid

Albite

Albite is a common felspar and is the "pivot" mineral of two different feldspar series. It is most often associated with the plagioclase series where it is an end member of this series. The plagioclase series comprises felspars that range in chemical composition from pure $NaAlSi_3O_8$ to pure $CaAl_2Si_2O_8$. The various plagioclase feldspars are identified from each other by gradations in index of refraction and density in the absence of chemical analysis and optical measurements. Albite is also an end member of the alkali or K-feldspars whose series ranges from pure $NaAlSi_3O_8$ to pure $KAlSi_3O_8$. This series only exists at high temperatures with the mineral sanidine being the potassium, K, rich end member. At lower temperatures, the K-feldspars will seperate from the albite in a process called exsolution. The albite will form layers inside the k-feldspars crystals. Some times these layers are discernable to the naked eye and the stone is referred to as perthite. Albite by definition must contain no less than 90% sodium and no more than 10% of either potassium and calcium in the cation position in the crystal structure.

Albite is the last of the feldspars to crystallize from molten rock. The process of crystallization from a molten rock body serves to isolate rarer elements in the last stages of crystallization and therefore produces rare mineral species. Thus albite is often found with some lovely rare and beautiful minerals. Although usually not an exceptional collection mineral in itself, albite can be a nice accessory mineral to other mineral species. A variety associated with tourmaline is called cleavelandite and forms extremely thin, platy, white and sometimes very transparent crystals.

All plagioclase feldspars show a type of twinning that is named after albite. Albite Law twinning produces stacks of twin layers that are typically only fractions of millimeters to several millimeters thick. These twinned layers can be seen as striation like grooves on the surface of the crystal and unlike true striations these also appear on the cleavage surfaces. The Carlsbad Law twin produces what appears to be two intergrown crystals growing in opposite directions. Two different twin laws, the Manebach and Baveno laws, produce crystals with one prominant mirror plane and penetrant angles or notches into the crystal. Although twinned crystals are common, single crystals showing a perfect twin are rare and are often collected by twin fanciers.

Physical Characteristics

- Color is usually white (Albite is derived from the same root word as albino) or colorless but can be shades of blue, yellow, orange and brown.
- Luster is vitreous to dull if weathered.
- Transparency crystals are translucent to opaque and only sometimes transparent.
- Crystal System is triclinic; bar 1.
- Crystal Habits include blocky, tabular and platy crystals. The typical crystal has a nearly rectangular or square cross-section with slightly slanted dome and pinacoid terminations. A variety called Cleavelandite forms very thin platy crystals that can grow rather large (15+ cm across) but can maintain an even thickness of only a few millimeters. Twinning is almost universal in albite. Crystals can be twinned according to the Albite, Carlsbad, Manebach and Baveno laws. Albite is a common constituent of granitic and syenite rocks. Can also be massive.
- Cleavage is perfect in one and good in another direction forming nearly right angled prisms.
- Fracture is conchoidal.
- Hardness is 6 - 6.5.
- Specific Gravity is approximately 2.61 (average).
- Streak is white.
- Associated Minerals are quartz, tourmaline and muscovite.
- Other Characteristics: index of refraction is 1.53. Lamellar twinning may cause a grooved effect on cystal surfaces that appear as striations. Some albite may show an opalescence due to twinning and is referred to as moonstone.
- Notable Occurrences include Labrador, Canada and the Scandinavian Peninsula.
- Best Field Indicators are occurence, crystal habit, twinning, striations, density and index of refraction.

Monoclinic Crystal System

Monoclinic system is one of the structural categories to which crystalline solids can be assigned. Crystals in this system are referred to three axes of unequal lengths—say, *a*, *b*, and *c*—of which *a* is perpendicular to *b* and *c*, but *b* and *c* are not perpendicular to each other.

If the atoms or atom groups in the solid are represented by points and the points are connected, the resulting lattice will consist of an orderly stacking of blocks, or unit cells. The monoclinic unit cell is distinguished by a single axis, called an axis of twofold symmetry, about which the cell can be rotated by 180° without changing its appearance. More solids belong to the monoclinic system than to any other. Beta-sulfur, gypsum, borax, orthoclase, kaolin, muscovite, clinoamphibole, clinopyroxene, jadeite, azurite, and spodumene crystallize in the monoclinic system.

Monoclinic crystal system

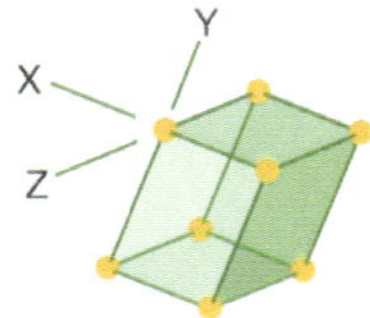

All three axes are unequal in length, and two axes are perpendicular to each other.

Its one variant is:

Base-centred monoclinic
lattice points in the middle of each of the two ends

Crystals in a monoclinic system are referred to three axes of unequal lengths, with two axes being perpendicular to each other.

In the monoclinic system all the axes are different lengths. Two of them, the A and the C axes, meet at 90 degrees but the third one does not.

This crystal system has the following forms:

- 3 Pinacoids
- Prism and 2 Pinacoids
- Prism and 3 Pinacoids
- Prism and 5 Pinacoids

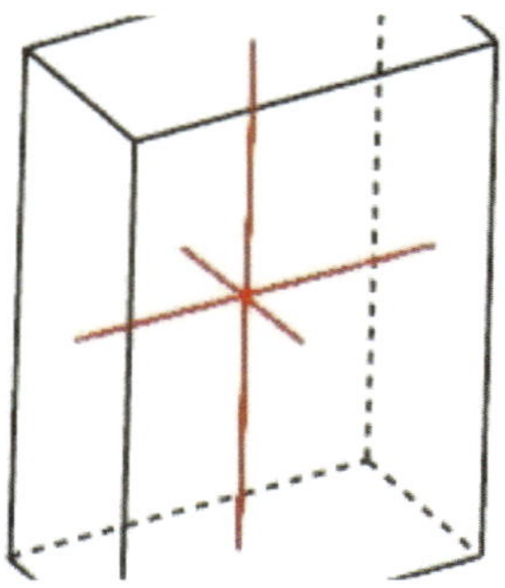
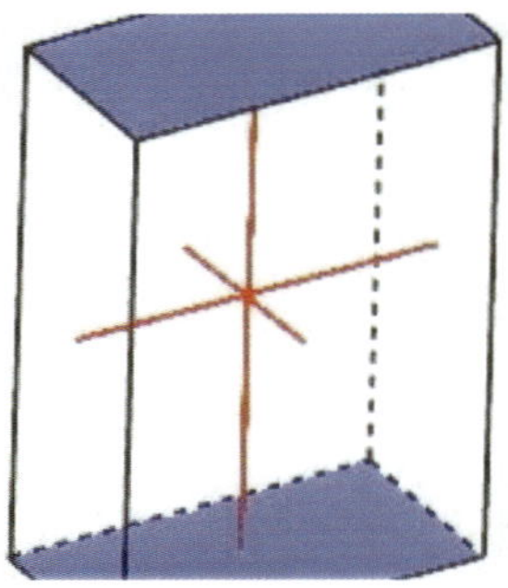

Annite

Annite is a phyllosilicate mineral in the mica family. It has a chemical formula of $KFe_3^{2+}AlSi_3O_{10}(OH)_2$. Annite is the iron end member of the biotite mica group, the iron-rich analog of magnesium-rich phlogopite.

Annite was first described in 1868 for the first noted occurrence in Cape Ann, Rockport, Essex County, Massachusetts, US. It also occurs on Pikes Peak, El Paso County, Colorado. It occurs in igneous and metamorphic rocks that are deficient in magnesium. It occurs associated with fluorite and zircon in the type locality.

General Information

- Category: Phyllosilicates (Mica group)
- Formula: $KFe_3^{2+}AlSi_3O_{10}(OH, F)_2$
- Crystal system: Monoclinic
- Crystal class: Prismatic (2/m).

Properties

Annite is monoclinic and contains tabular crystals and cleavage fragments with pseudohexagonal outlines. There are contact twins with composition surface {001} and twin axis {310}.

- Color: Brown to black.
- Crystal habit: usually in “blocks” of thin sheets or as aggregations of foliated thin scales.

- Twinning: contact twins with composition surface on {001} and twin axis {310}.
- Fracture: uneven.
- Mohs scale hardness: 2.5 – 3.
- Luster: submetallic to vitreous; pearly on cleavage surfaces.
- Streak: brownish white.
- Diaphaneity: Translucent to transparent.
- Specific gravity: approximately 3.3.

The relief of a mineral refers to the way a mineral may stand out in the plane polarized light. A mineral may be referred to as having a low or high relief. Minerals with a high relief, such as annite, have sharp grain boundaries and display good fracture and cleavage. When viewed under a microscope, this mineral may appear to stick out of the other minerals in the thin section. Relief primarily depends on the index of refraction of the mineral. The index of refraction of a mineral is a measure of the speed of light in the mineral. It is expressed as a ratio of the speed of light in vacuum relative to that in the given mineral.

Occurrence and uses

Annite is a member of the mica group and has very similar properties as other micas such as muscovite and biotite. More importantly, annite is interesting to geologists because it can be used for potassium-argon dating. Because annite contains large amounts of potassium, it can be used to find the absolute age of articles older than 1000 years. This type of dating also preserves a record of the direction and intensity of the local magnetic field, giving field geologists better knowledge of their surroundings.

Orthorhombic Crystal System

The orthorhombic crystal system includes all crystals with a rhombic inner structure. The word ‹rhombic' is derived from the Greek *rhombus*. Rhombic minerals form either rhombic crystals, for example aragonite,olivine (peridot) or topaz, but sometimes also hexagonal ones, as three rhombuses fitted together form a hexagon. This type of growth is quite aptly termed 'pseudohexagonal'; an example is aragonite.

In this system there are three axes, all of which meet at 90 degrees to each other. However, all the axes are a different length.

This crystal system has the following forms:

- Prism and Pinacoid
- Three Prism
- 2 Pinacoid Form

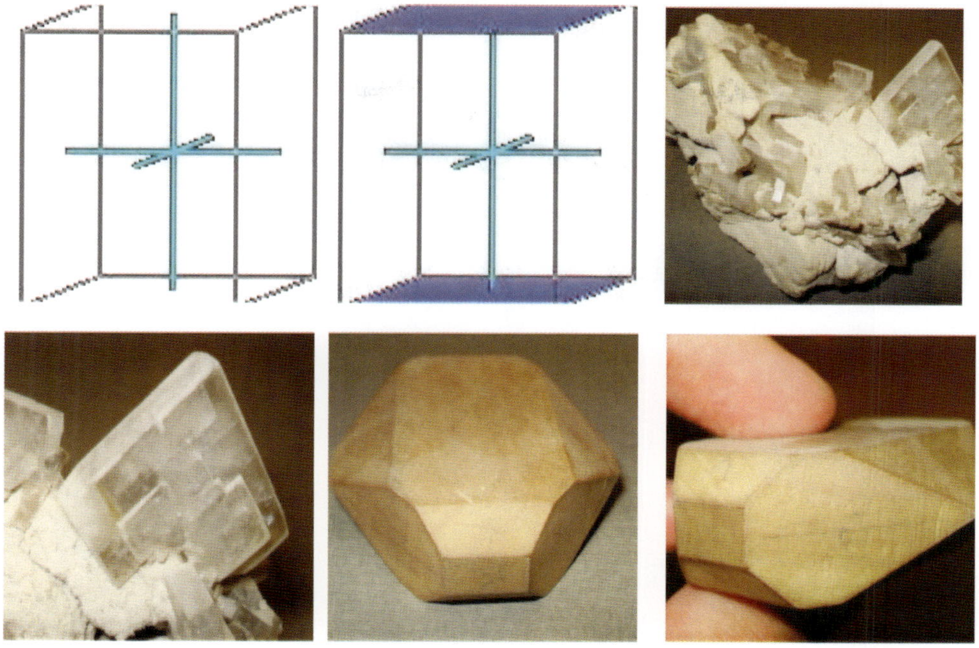

Epsomite

Epsomite is a hydrous magnesium sulfate mineral with formula $MgSO_4{\cdot}7H_2O$. It was first systematically described in 1806 for an occurrence near Epsom, Surrey, England, after which it was named. It is actually well known in most households as the artificially created epsom salt. It absorbs water from the air and converts to hexahydrate with the loss of one water molecule and a switch to monoclinic structure.

General Information

- Category: Sulfate mineral
- Formula: $MgSO_4 7H_2O$
- Crystal system: Orthorhombic
- Crystal class: Disphenoidal (222)

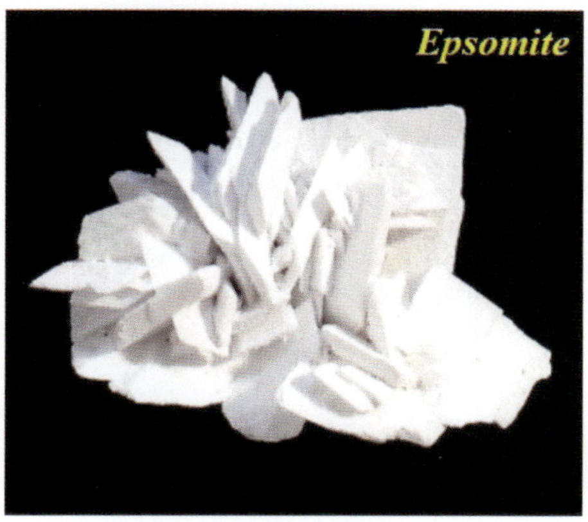

Properties

Epsomite crystallizes in the orthorhombic system as rarely found acicular or fibrous crystals, the normal form is as massive encrustations. It is colorless to white with tints of yellow, green and pink. The Mohs hardness is 2 to 2.5 and it has a low specific gravity of 1.67.

- Color: White, grey, colorless, or pink, greenish
- Crystal habit: Acicular to fibrous encrustations
- Fracture: Conchoidal
- Mohs scale hardness: 2
- Luster: Vitreous, silky when fibrous
- Diaphaneity: Transparent to translucent
- Specific gravity: 1.67 – 1.68

Discovery and Occurrence

Epsomite forms from as an efflorescent (a precipitation from vapors), on limestone cave walls and on the walls and timbers of various mines. It has been found in deposits from hot springs and fumaroles such as on Mt. Vesuvius, Italy. It forms as encrustations or efflorescences on limestone cavern walls and mine timbers and walls, rarely as volcanic fumarole deposits, and as rare beds in evaporite layers.

It occurs in association with melanterite, gypsum, halotrichite, pickeringite, alunogen, rozenite, and mirabilite.

Tetragonal Crystal System

Tetragonal Crystal Systems include all crystals with a rectangular inner structure. Tetragonal minerals usually form crystals with rectangualr pillars. They sometimes have flat boundaires but can also have pyramid points. Some examples would include Apophyllite, Rutile and Zircon.

Like the cubic system, the tetragonal system has three axes that all meet at 90 degrees. It differs from the isometric system in that the C axis is longer than the A and B axis which are the same length. This makes it's for a rectangular prism. It is important to note that the tetragonal system will have a square base and top, but its height will be longer than the axes which make the square base/top.

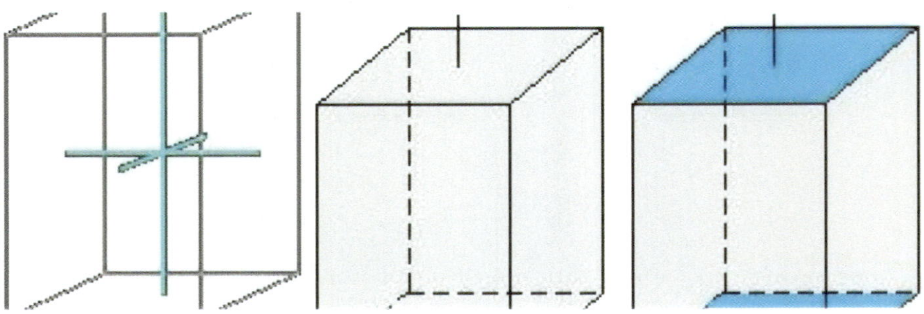

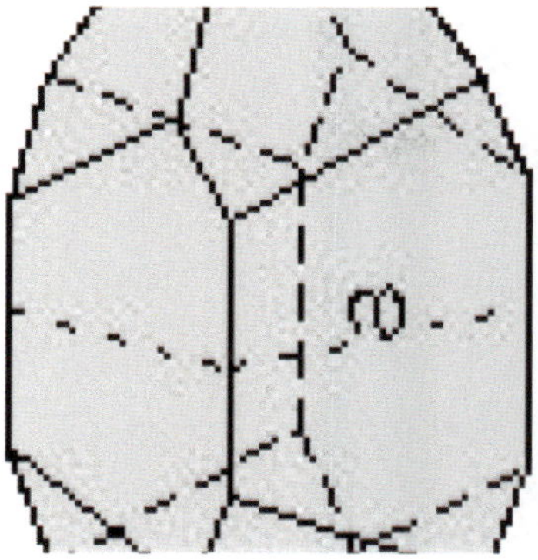

Wulfenite

Wulfenite is a nice collection type mineral and is popular as such. Its strong colors, nice luster and one-of-a-kind crystal habits attract the attention of many collectors around the world.

Wulfenite is an enigma in terms of its symmetry. There are conflicting results of various symmetry tests and this usually does not happen. It is either a symmetry of 4 or 4/m. The difference is the disputed existence of a mirror plane perpendicular to the four fold axis. If the mirror exists, then the crystals should have a top that is a mirror image of its bottom. Although most crystals don't show it clearly, the bottom pyramidal faces slant at a different angle from the top pyramidal faces. This demonstrates the symmetry of just 4. However, other tests of its symmetry show a 4/m symmetry. This symmetrical oddity only adds to wulfenite's interest among serious collectors.

Physical Characteristics

- Color is red, orange, yellow, silver and white.
- Luster is vitreous.
- Transparency: Crystals are transparent to translucent.
- Crystal System is tetragonal; 4/m or 4.
- Crystal Habits include very thin square or octahedral pinacoidal plates with pyramidal faces truncating just the edges of the crystal. At times the pyramids become prominant and psuedo-dipyramidal crystal habits are seen, sometimes because of twinning. Prismatic faces are also seen and can make psuedo-cubic crystals. Also encrusting and cavernous aggregates due to intergrowth of crystal plates.
- Cleavage is perfect in one direction.
- Fracture is conchoidal.
- Hardness is 3.
- Specific Gravity is approximately 6.8 (very heavy for translucent minerals).
- Streak is white.

- Associated. Minerals are mimetite, limonite, smithsonite, vanadinite and galena.
- Other Characteristics: index of refraction is 2.28-2.40 (very high, but typical of lead minerals).
- Notable Occurances include Morocco; Tsumeb, Nambia; Mexico and Arizona and New Mexico, USA.
- Best Field Indicators are crystal habit, color, density and luster.

Chalcopyrite

Auriferous Chalcopyrite: A specimen of chalcopyrite with pyrrhotite from the Rouyn District, Quebec, Canada. Some chalcopyrite contains enough gold or silver that it can be an ore of those metals without considering the copper content. This specimen is about ten centimeters across

Chalcopyrite is a brass-yellow mineral with a chemical composition of $CuFeS_2$. It occurs in most sulfide mineral deposits throughout the world and has been the most important ore of copper for thousands of years.

The surface of chalcopyrite loses its metallic luster and brass-yellow color upon weathering. It tarnishes to a dull, gray-green color, but in the presence of acids the tarnish can develop a red to blue to purple iridescence.

The iridescent colors of weathered chalcopyrite attract attention. Some souvenir shops sell chalcopyrite that has been treated with acid as "peacock ore." But, "peacock ore" is a more appropriate name for the mineral bornite.

Treated Chalcopyrite: These specimens of chalcopyrite were treated with acid to give them a blue and purple iridescence, which enhances their visual appeal and marketability.

Physical Properties of Chalcopyrite

The most obvious physical properties of chalcopyrite are its brassy yellow color, metallic luster, and high specific gravity. These give it a similar appearance to pyrite and gold. Distinguishing these minerals is easy. Gold is soft, has a yellow streak and has a much higher specific gravity. Chalcopyrite is brittle and has a greenish gray streak. Pyrite is hard enough that it cannot be scratched with a nail, but chalcopyrite is easily scratched with a nail.

The name "fool's gold" is most often associated with pyrite because it is more common and more often confused with gold. Chalcopyrite is also confused with gold, so the name "fool's gold" is also applied and appropriate.

Chalcopyrite on Dolomite: Tetragonal crystals of chalcopyrite on dolomite from Baxter Springs, Kansas. This specimen is about 10 centimeters across.

Physical Properties of Chalcopyrite	
Chemical Classification	Sulfide
Color	Brass yellow. Tarnishes to gray green, sometimes iridescent.
Streak	Greenish black
Luster	Metallic
Diaphaneity	Opaque
Cleavage	Poor
Mohs Hardness	3.5 to 4
Specific Gravity	4.1 to 4.3
Diagnostic Properties	Color, greenish streak, softer than pyrite, brittle.
Chemical Composition	Copper iron sulfide, $CuFeS_2$.
Crystal System	Tetragonal
Uses	The most important ore of copper for thousands of years.

Chalcopyrite: Chalcopyrite from Ajo, Arizona. Specimen is approximately 10 centimeters across.

Chalcopyrite: Specimen of chalcopyrite from Rouyn District, Quebec, Canada.
Specimen is approximately 10 centimeters across.

Geologic Occurrence of Chalcopyrite

Chalcopyrite forms under a variety of conditions. Some is primary, crystallizing from melts as accessory minerals in igneous rocks. Some forms by magmatic segregation and is in the stratified rocks of a magma chamber. Some occurs in pegmatite dikes and contact metamorphic rocks. Some is disseminated through schist and gneiss. Many volcanogenic massive sulfide deposits containing chalcopyrite are known.

The most significant chalcopyrite deposits to be mined are hydrothermal in origin. In these, some chalcopyrite occurs in veins and some replaces country rock. Associated ore minerals include pyrite, sphalerite, bornite, galena, and chalcocite.

Chalcopyrite serves as the copper source for many secondary mineral deposits. Copper is removed from chalcopyrite by weathering or solution, transported a short distance, then redeposited as secondary sulfide, oxide, or carbonate minerals. Many malachite, azurite, covellite, chalcocite, and cuprite deposits contain this secondary copper.

Uses of Chalcopyrite

The only important use of chalcopyrite is as an ore of copper, but this single use should not be understated. Chalcopyrite has been the primary ore of copper since smelting began over five thousand years ago.

Some chalcopyrite ores contain significant amounts of zinc substituting for iron. Others contain enough silver or gold that the precious metal content more than pays the costs of mining.

Leucite

Leucite is a popular and interesting mineral. Its name comes from the greek word for "white" in allusion to its typical color. At high temperatures, leucite is isometric and will form the isometric trapezohedron crystal form. Interestingly, as leucite cools, the isometric structure becomes unstable and transforms into a tetragonal structure without altering the outward shape. Although the mineral is actually tetragonal, the outward shape is pseudo-isometric and thus the crystal form is actually pseudo-trapezohedral.

Leucite is one of the few minerals that forms the unique trapezohedron. The trapezohedron has 24 deltoid shaped faces, where each face occupies one third of the position of a single octahedron's face. The minerals of the garnet group and the mineral analcime are the only common minerals that will also form the trapezohedron.

Distinguishing leucite from the garnets and analcime is relatively easy in some cases. The garnets are much harder and usually deeply colored. Leucite has a much lower density and usually has a duller luster than analcime. Also leucite is typically embedded in host rock where as analcime, when displaying good crystals and not massive or granular, is loose or attacted to other minerals in volcanic cavities.

Leucite, $KAlSi_2O_6$ is actually distantly related to analcime, $NaAlSi_2O_6$-H_2O. Leucite is a member of the feldspathoid groupof minerals. Analcime, although usually considered a zeolite, is sometimes placed in the feldspathoid group since its chemistry and occassional occurrences are similar.

Minerals whose chemistries are close to that of the alkali feldspars but are poor in silica (SiO2) content, are called feldspathoids. Leucite, like other feldspathoids, is found in silica poor rocks containing other silica poor minerals and no quartz. If quartz were present when the melt was crystallizing, it would react with any feldspathoids and form a feldspar.

At one time leucite was used as a source of potassium and aluminum. Probably due to the high aluminum to silicon ratio, its structure is easily destroyed by acids and this frees the aluminum ions.

Leucite, already a pseudomorph from a high temperature phase, commonly is altered to pseudoleucite. Pseudoleucite is not a mineral but a mixture of nepheline, orthoclase and analcime.

Physical Characteristics

- Color is clear, white or gray, with yellowish and reddish tints possible.
- Luster is vitreous or greasy to dull.
- Transparency: crystals are transparent, translucent to commonly opaque.
- Crystal System is isometric; 4/m bar 3 2/m at temperatures above 605 degrees C, and tetragonal; either 4 2 2 or 4/m, below 605 degrees C.
- Crystal Habits include the characteristic trapezohedron (actually pseudo-trapezohedron). Also granular and massive.
- Cleavage is absent.
- Fracture is conchoidal.
- Hardness is 5.5 - 6.
- Specific Gravity is approximately 2.4 - 2.5 (slightly below average)
- Streak is white.
- Associated Minerals include olivine, labradorite, augite, biotite, nepheline and other feldspathoids.
- Other Characteristics: surface is usually pitted, dull or has a weathered look.

- Notable Occurrences include Mt. Vesuvius, Italy; Magnet Cove, Arkansas, Leucite Hills, Wyoming and Litchfield, Maine, USA; Brazil and Hastings Co, Ontario.
- Best Field Indicators are crystal habit, density, low hardness, luster and associations.

Hexagonal Crystal Family

The hexagonal system has four crystallographic axes consisting of three equal horizontal, or equilateral axes at 120 degrees to each other, as well as one vertical axis which is perpendicular to the other three. This vertical axis can be longer or shorter than the horizontal axes.

The system has the following forms:

- Prism
- Pinacoid
- Dipyramid
- Ditrigonal Pyramid
- Trigonal Prism
- Ditrigonal Prism.

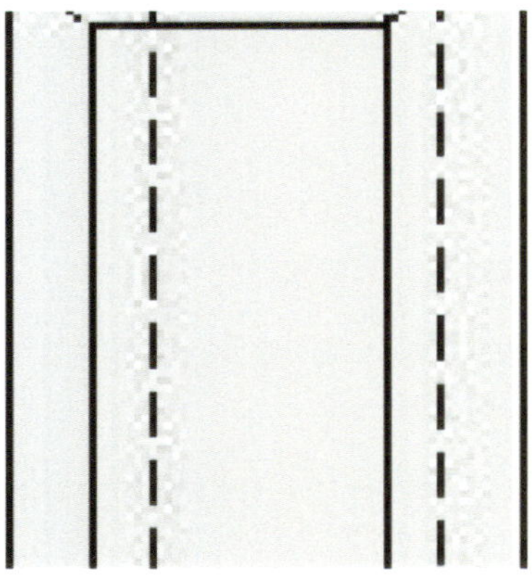

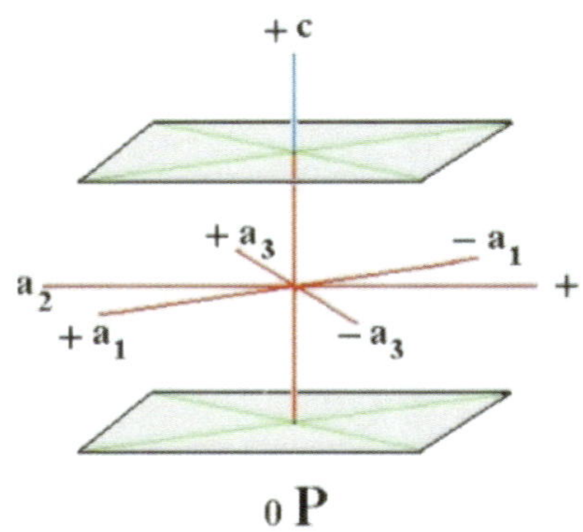

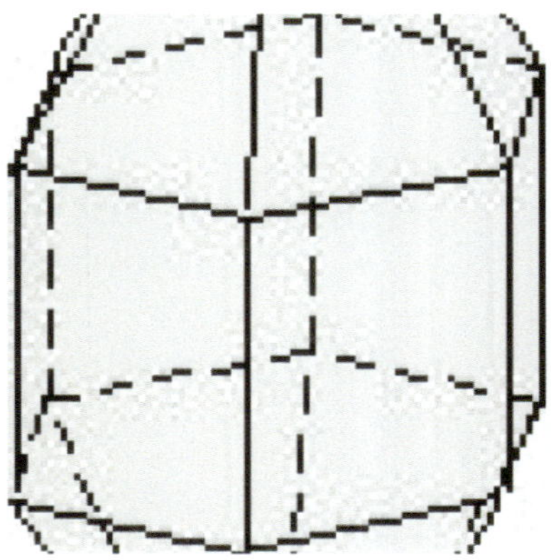

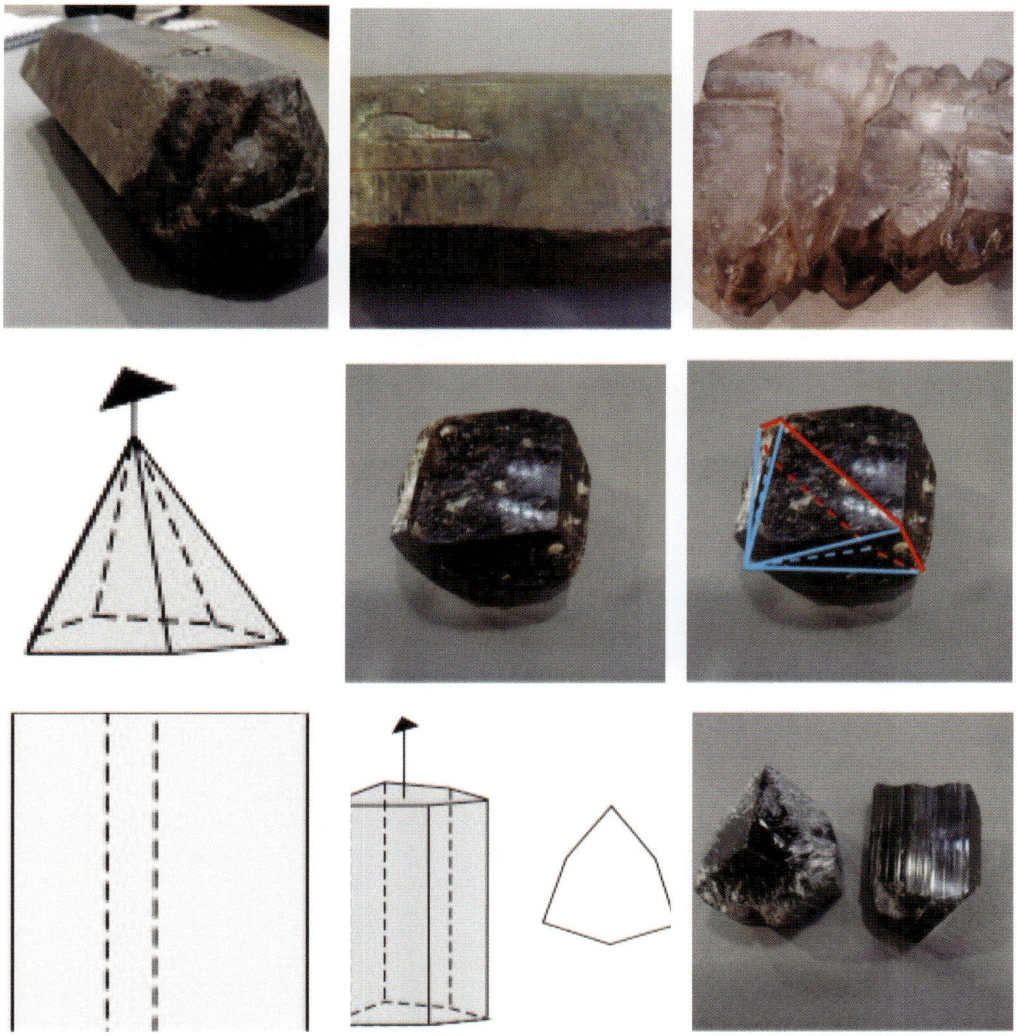

Graphite

Graphite is a naturally-occurring form of crystalline carbon. It is a native element mineral found in metamorphic and igneous rocks. Graphite is a mineral of extremes. It is extremely soft, cleaves with very light pressure, and has a very low specific gravity. In contrast, it is extremely resistant to heat and nearly inert in contact with almost any other material. These extreme properties give it a wide range of uses in metallurgy and manufacturing.

Flake graphite: Flake graphite produced in Madagascar.

Graphite chunk: Lump graphite from Kropfmuhl, Austria. Specimen is about one and one half inches (3.8 cm) across.

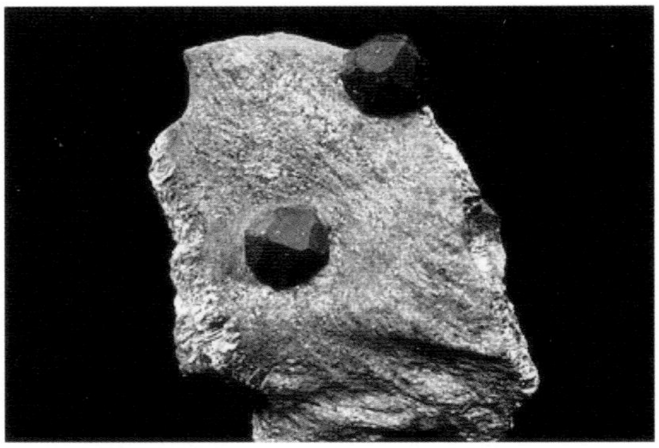

Graphite with garnet: A specimen of graphite-mica schist with two red almandine/pyrope garnets from the Red Embers Mine, Erving, Massachusetts. This specimen is about two inches (5.08 cm) across.

Geologic Occurrence

Graphite is a mineral that forms when carbon is subjected to heat and pressure in Earth's crust and in the upper mantle. Pressures in the range of 75,000 pounds per square inch and temperatures in the range of 750 degrees Celsius are needed to produce graphite. These correspond to the granulite metamorphic facies.

Graphite From Regional Metamorphism (Flake Graphite)

Most of the graphite seen at Earth's surface today was formed at convergent plate boundaries where organic-rich shales and limestones were subjected to the heat and pressure of regional metamorphism. This produces marble, schist, and gneiss that contain tiny crystals and flakes of graphite.

When graphite is in high enough concentrations, these rocks can be mined, crushed to a particle size that liberates the graphite flakes, and processed by specific gravity separation or froth flotation to remove the low-density graphite. The product produced is known as "flake graphite."

Graphite from Coal Seam Metamorphism ("Amorphous" Graphite)

Some graphite forms from the metamorphism of coal seams. The organic material in coal is composed mainly of carbon, oxygen, hydrogen, nitrogen, and sulfur. The heat of metamorphism destroys the organic molecules of coal, volatilizing the oxygen, hydrogen, nitrogen, and sulfur. What remains is a nearly pure carbon material that crystallizes into mineral graphite.

This graphite occurs in "seams" that correspond to the original layer of coal. When mined, the material is known as "amorphous graphite." The word "amorphous" is actually incorrect in this usage, as it does have a crystalline structure. From the mine, this material has an appearance similar to lumps of coal without the bright and dull banding.

Graphite from Hydrothermal Metamorphism

A small amount of graphite forms by the reaction of carbon compounds in the rock during hydrothermal metamorphism. This carbon can be mobilized and deposited in veins in association with hydrothermal minerals. Because it is precipitated, it has a high degree of crystallinity, and that makes it a preferred material for many electrical uses.

Graphite in Igneous Rocks and Meteorites

Small amounts of graphite are known to occur as a primary mineral in igneous rocks. It is known as tiny particles in basalt flows and syenite. It is also known to form in pegmatite. Some iron meteorites contain small amounts of graphite. These forms of graphite are occurrences without economic importance.

Synthetic Graphite

"Synthetic graphite" is made by heating high-carbon materials like petroleum coke and coal-tar pitch to temperatures in the range of 2500 to 3000 degrees Celsius. At these high temperatures, all volatile materials and many metals in the feedstock are destroyed or driven off. The graphite that remains links into a sheet-like crystalline structure. Synthetic graphite can have a purity of over 99% carbon, and it is used in manufactured products where an extremely pure material is required.

Cubic Crystal System

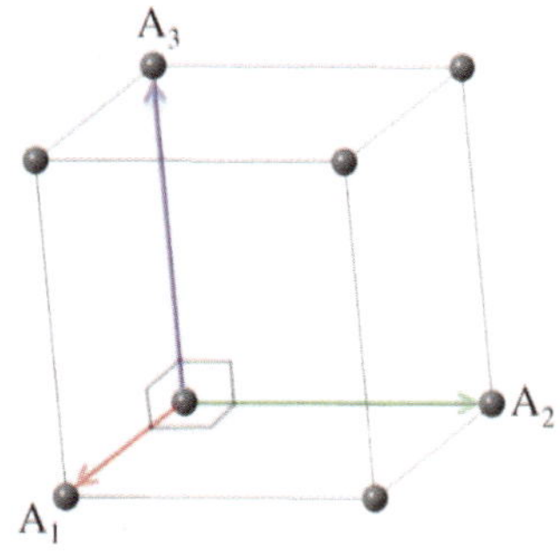

The cubic crystal system is defined as having the symmetry of a cube: the conventional unit cell can be rotated by 90° about any axis, or by 180° around an axis running through the center of two opposing cube edges, or by 120°around a body diagonal, and retain the same shape. The conventional cell then takes the form,

$$A_1 = a\hat{x}$$
$$A_2 = a\,\hat{y}$$
$$A_3 = a\,\hat{z},$$

with unit cell volume,

$$V = a^3 .$$

This is the limiting case of both the orthorhombic and tetragonal systems when all primitive vectors are equal in length. There are three Bravais lattices in the cubic system.

Lattice 1: Simple Cubic

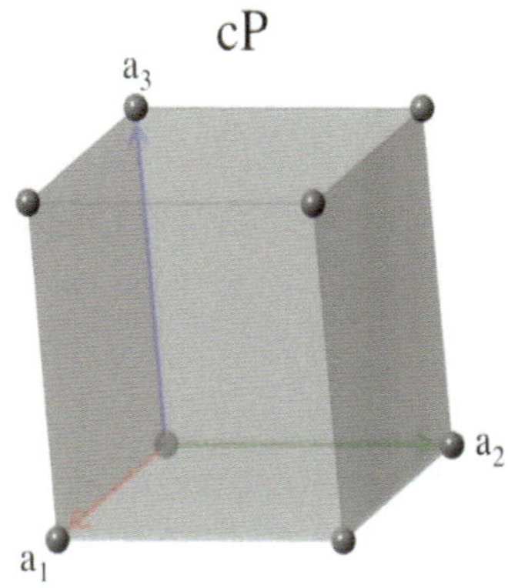

The simple cubic system is identical to the conventional cubic unit cell,

$$a_1 = a\hat{x}$$
$$a_2 = a\,\hat{y}$$
$$a_3 = a\,\hat{z},$$

with volume,

$$V = a^3 .$$

This can also be considered as a rhombohedral lattice with $\alpha = \pi/2$. The space groups associated with this lattice are

195. P23	198. $P2_13$	200. $Pm\overline{3}$
201. $Pn\overline{3}$	205. $pa\overline{3}$	207. p432
208. $P4_232$	212. $p4_332$	213. $p4_132$
215. $P\overline{4}3m$	218. $P\overline{4}3n$	221. $Pm\overline{3}m$
222. $Pn\overline{3}n$	223. $Pm\overline{3}n$	224. $Pn\overline{3}m$

Lattice 2: Face-Centered Cubic

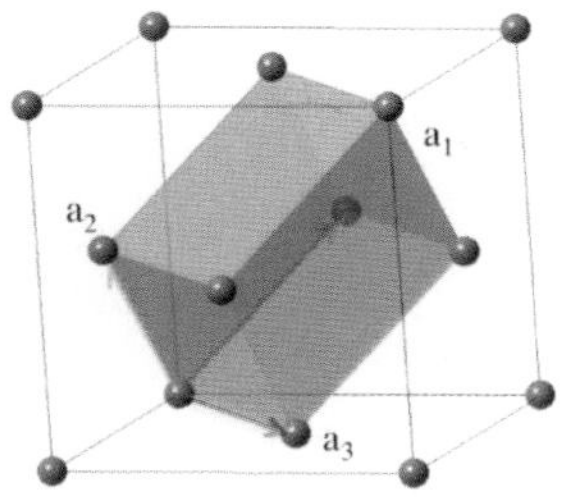

The face-centered cubic lattice can be derived from its predecessors in the orthorhombic and tetragonal systems, having the same periodicity as its simple cubic parent with the addition of a translation from one corner of the cube to the center of any face. Our standard face-centered cubic primitive vectors have the form,

$$a_1 = \frac{a}{2}\hat{y} + \frac{a}{2}\hat{z}$$

$$a_2 = \frac{a}{2}\hat{x} + \frac{a}{2}\hat{z}$$

$$a_3 = \frac{a}{2}\hat{x} + \frac{a}{2}\hat{y},$$

and the primitive cell volume is,

$$V = \frac{a^3}{4}.$$

There are four face-centered cubic primitive cells in the conventional cubic cell. The face-centered cubic lattice can be considered as a rhombohedral lattice where $\alpha = 60^\circ$. The space groups associated with this lattice are,

196. $F23$	202. $Fm\bar{3}$	203. $Fd\bar{3}$
209. $F432$	210. $F4_1 32$	216. $F\bar{4}3m$
219. $F\bar{4}3c$	225. $Fm\bar{3}m$	226. $Fm\bar{3}c$
227. $Fd\bar{3}m$	228. $Fd\bar{3}c$	

Lattice 3: Body-Centered Cubic

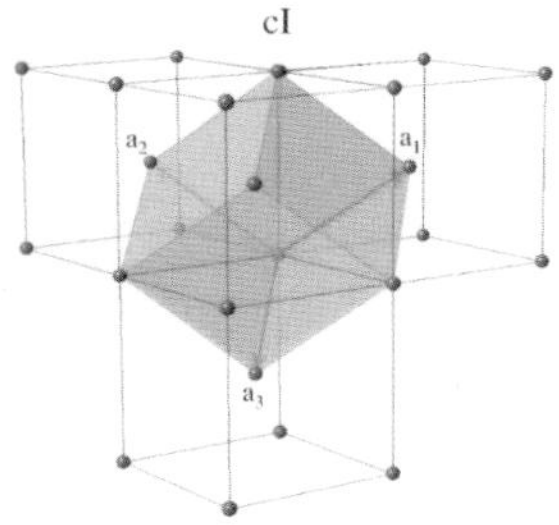

Like its predecessors in the orthorhombic and tetragonal systems, the body-centered cubic crystal has the same periodicity as its parent with the addition of a translation from one corner of the cube to its center. Our standard body-centered cubic primitive vectors have the form,

$$\mathrm{a}_1 = -\frac{a}{2}\hat{\mathrm{x}} + \frac{a}{2}\hat{\mathrm{y}} + \frac{a}{2}\hat{\mathrm{z}}$$

$$\mathrm{a}_2 = \frac{a}{2}\hat{\mathrm{x}} - \frac{a}{2}\hat{\mathrm{y}} + \frac{a}{2}\hat{\mathrm{z}}$$

$$\mathrm{a}_3 = \frac{a}{2}\hat{\mathrm{x}} + \frac{a}{2}\hat{\mathrm{y}} - \frac{a}{2}\hat{\mathrm{z}},$$

and the primitive cell volume is

$$V = \frac{a^3}{2}.$$

There are two body-centered cubic primitive cells in the conventional cubic cell. The body-centered cubic lattice can be considered as a rhombohedral lattice where $\alpha = \cos^{-1}(-1/3) \approx 109.47^\circ$. The space groups associated with this lattice are

197. I23	199. $I2_13$	204. $Im\bar{3}$
206. $Ia\bar{3}$	211. I432	214. $I4_132$
217. $I\bar{4}3m$	220. $I\bar{4}3d$	229. $Im\bar{3}m$
230. $Ia\bar{3}d$		

Sphalerite

Sphalerite with galena and chalcopyrite: A typical mineral association of sphalerite with galena and chalcopyrite. From the Huaron Mine of Peru. Specimen is about 4.3 x 3.2 x 1.8 centimeters in size.

Sphalerite is a zinc sulfide mineral with a chemical composition of (Zn,Fe)S. It is found in metamorphic, igneous, and sedimentary rocks in many parts of the world. Sphalerite is the most commonly encountered zinc mineral and the world's most important ore of zinc.

Dozens of countries have mines that produce sphalerite. Recent top producers include Australia, Bolivia, Canada, China, India, Ireland, Kazakhstan, Mexico, Peru, and the United States. In the United States, sphalerite is produced in Alaska, Idaho, Missouri, and Tennessee.

The name sphalerite is from the Greek word "sphaleros" which means deceiving or treacherous. This name is in response to the many different appearances of sphalerite and because it can be challenging to identify in hand specimens. Names for sphalerite used in the past or by miners include "zinc blende," "blackjack," "steel jack," and "rosin jack."

Geologic Occurrence

Many minable deposits of sphalerite are found where hydrothermal activity or contact metamorphism has brought hot, acidic, zinc-bearing fluids in contact with carbonate rocks. There, sphalerite can be deposited in veins, fractures, and cavities, or it can form as mineralizations or replacements of its host rocks.

In these deposits, sphalerite is frequently associated with galena, dolomite, calcite, chalcopyrite, pyrite, marcasite, and pyrrhotite. When weathered, the zinc often forms nearby occurrences of smithsonite or hemimorphite.

Sphalerite on dolomite: Crystals of sphalerite on dolomite with minor amounts of chalcopyrite. Specimen from the Joplin Field, Tri-State District of Missouri, USA. Specimen is about 6.5 x 4.5 x 3.5 centimeters.

Chemical Composition

The chemical formula of sphalerite is (Zn,Fe)S. It is a zinc sulfide containing variable amounts of iron that substitutes for zinc in the mineral lattice. The iron content is normally less than 25% by weight. The amount of iron substitution that occurs depends upon iron availability and temperature, with higher temperatures favoring higher iron content.

Sphalerite often contains trace to minor amounts of cadmium, indium, germanium, or gallium. These rare elements are valuable and when abundant enough can be recovered as profitable byproducts. Minor amounts of manganese and arsenic can also be present in sphalerite.

Sphalerite crystals: Gem-quality crystals of yellow sphalerite from the Balmat-Edwards Zinc District of New York. Specimen is about 2.75 x 1.75 x 1.5 centimeters in size.

Sphalerite: Sphalerite with dolomite from Gilman, Colorado.
Specimen is approximately 5 centimeters across.

Physical Properties

The appearance and properties of sphalerite are variable. It occurs in a variety of colors, and its luster ranges from nonmetallic to submetallic and resinous to adamantine. Occasionally it will be transparent with a vitreous luster. Sphalerite's streak is white to yellowish brown and sometimes is accompanied by a distinct odor of sulfur. Occasionally it streaks reddish brown.

One of the most distinctive properties of sphalerite is its cleavage. It has six directions of perfect cleavage with faces that exhibit a resinous to adamantine luster. Specimens that display this distinctive cleavage are easy to identify. Unfortunately, many specimens have such a fine grain size that the cleavage is difficult to observe.

Because sphalerite often forms in veins and cavities, excellent crystals are relatively common. Sphalerite is a member of the isometric crystal system, and cubes, octahedrons, tetrahedrons, and dodecahedrons are all encountered.

Physical Properties of Sphalerite	
Chemical Classification	Sulfide
Color	Yellow, brown, black, red, green, white, colorless
Streak	White to yellowish brown, often with an odor of sulfur
Luster	Nonmetallic, submetallic, resinous or adamantine
Diaphaneity	Transparent to translucent
Cleavage	Perfect, dodecahedral, in six directions!
Mohs Hardness	3.5 to 4
Specific Gravity	3.9 to 4.1
Diagnostic Properties	Luster, cleavage, streak
Chemical Composition	Zinc sulfide with variable amounts of iron, (Zn,Fe)S
Crystal System	Isometric
Uses	The primary ore of zinc. Often mined for minor amounts of indium, cadmium, germanium, or gallium as profitable byproducts. Mineral specimens. Faceted stones for collectors.

Sphalerite gemstones: Sphalerite is occasionally cut as a faceted stone. It is a popular stone with collectors because it has a dispersion that is three times higher than the dispersion of diamond. The stones occur in a spectrum of colors ranging from yellowish green, to yellow, to orange, to red. To exhibit the excellent dispersion the stones must have a very high clarity. Sphalerite's very low hardness (3.5 to 4 on the Mohs scale) and perfect cleavage make the mineral a very poor choice for any jewelry except pieces such as earrings and brooches that will receive very little abrasion or impact. It is considered to be a "collector's stone."

Sphalerite as a Gemstone

Although sphalerite has a hardness of just 3.5 to 4 on the Mohs scale and is not suitable for most jewelry use, specimens with excellent clarity are sometimes cut into gemstones for collectors. Why? Sphalerite has a dispersion that exceeds that of all of the popular gems and is three times higher than the dispersion of diamond.

Dispersion is the ability of a material to separate white light into the colors of the spectrum as it passes through the material. Diamond is well known for its exceptional "fire" - flashes of color as the gem is moved under a source of light. These are caused by its high dispersion of 0.044. Common natural gems that have a dispersion higher than diamond are sphene at 0.051 and demantoid garnet at 0.057. Sphalerite has an incredible dispersion of 0.156. The only things that hold specimens of sphalerite back from an incredible display of brilliant "fire" are their less-than-excellent clarity and their obvious body color.

Sphalerite is a difficult stone to cut and polish. It is soft and it has cleavage. Weaknesses in the stone or minor accidents during the cutting or polishing process can easily ruin a stone. Before deciding to cut a transparent specimen of sphalerite into a gemstone, its value as a mineral specimen should be determined. If this is not done, the owner could make a costly mistake.

Pyrite

Pyrite is the classic "Fool's Gold". There are other shiny brassy yellow minerals, but pyrite is by far the most common and the most often mistaken for gold. Whether it is the golden look or something else, pyrite is a favorite among rock collectors. It can have a beautiful luster and interesting crystals. It is so common in the earth's crust that it is found in almost every possible environment, hence it has a vast number of forms and varieties.

Bravoite is the name given to a nickel-rich iron sulfide. It is closely related to pyrite but contains up to 20% nickel. Some mineral books treat it as a variety of pyrite.

Pyrite is a polymorph of marcasite, which means that it has the same chemistry, FeS_2, as marcasite; but a different structure and therefore different symmetry and crystal shapes. Pyrite is difficult to distinguish from marcasite when a lack of clear indicators exists.

Pyrite's structure is analogous to galena's structure with a formula of PbS. Galena, though, has a higher symmetry. The difference between the two structures is that the single sulfur of galena is replaced by a pair of sulfurs in pyrite. The sulfur pair are covalently bonded together in essentially an elemental bond. This pair disrupts the four fold symmetry that a single atom of sulfur would have preserved and thus gives pyrite a lower symmetry than galena.

Although pyrite is common and contains a high percentage of iron, it has never been used as a significant source of iron. Iron oxides such as hematite and magnetite, are the primary iron ores. Pyrite is not as ecomonical as these ores possibly due to their tendency to form larger concentrations of more easily mined material. Pyrite would be a potential source of iron if these ores should become scarce.

Pyrite has been mined for its sulfur content, however. During WW II, sulfur was in demand as a strategic chemical and North American native sulfur mines were drying up. A sulfide deposit near Ducktown, Tennessee contained commercially valuable deposits of pyrite and other sulfides such as pyrrhotite and pentlandite and produced the needed sulfur as well as iron and other metals. The sulfur was used in the production of sulfuric acid, an important chemical for industrial purposes. Now most sulfur production comes from H_2S gas recovered from natural gas wells.

Physical Characteristics

- Color is brassy yellow.
- Luster is metallic.
- Transparency: Crystals are opaque.
- Crystal System is isometric; bar 3 2/m.
- Crystal Habits include the cube, octahedron and pyritohedron (a dodecahedron with pentagonal faces) and crystals with combinations of these forms. Good interpenetration twins called iron crosses are rare. Pyrite is commonly found in nodules. A flattened nodular variety called "*Pyrite Suns*" or "*Pyrite Dollars*" is popular in rock shops. Also massive or reniform and replaces other minerals and fossils forming pseudomorphs or copies.
- Cleavage is very indistinct.
- Fracture is conchoidal.
- Hardness is 6 - 6.5.
- Specific Gravity is approximately 5.1+ (heavier than average for metallic minerals).
- Streak is greenish black.
- Other Characteristics: Brittle, striations on cubic faces caused by crossing of pyritohedron

with cube. (note - striations on cube faces also demonstrate pyrite's lower symmetry). Pyrite (unlike gold) is not malleable.

- Associated Minerals are quartz, calcite, gold, sphalerite, galena, fluorite and many other minerals. Pyrite is so common it may be quicker to name the unassociated minerals.
- Notable Occurrences include Illinois and Missouri, USA; Peru; Germany; Russia; Spain; and South Africa among many others.
- Best Field Indicators are crystal habit, hardness, streak, luster and brittleness.

Lattice Constant

A lattice constant describes the spacing between adjacent unit cells in a crystal structure. The unit cells or building blocks of the crystal are three dimensional and have three linear constants that describe the cell dimensions. The dimensions of the unit cell are determined by the number of atoms packed into each cell and by how the atoms are arranged. A hard-sphere model is adopted, which allows you to visualize atoms in the cells as solid spheres. For cubic crystal systems, all three linear parameters are identical, so a single lattice constant is used to describe a cubic unit cell.

Identify the Space Lattice

Identify the space lattice of the cubic crystal system based on the arrangement of the atoms in the unit cell. The space lattice may be simple cubic (SC) with atoms only positioned at the corners of the cubic unit cell, face-centered cubic (FCC) with atoms also centered in every unit cell face, or body-centered cubic (BCC) with an atom included in the center of the cubic unit cell. For example, copper crystallizes in an FCC structure, while iron crystallizes in a BCC structure. Polonium is an example of a metal that crystallizes in a SC structure.

Find the Atomic Radii

Find the atomic radius (r) of the atoms in the unit cell. A periodic table is an appropriate source for atomic radii. For example, the atomic radius of polonium is 0.167 nm. The atomic radius of copper is 0.128 nm, while that of iron is 0.124 nm.

Calculate the Lattice Constant

Calculate the lattice constant, a, of the cubic unit cell. If the space lattice is SC, the lattice constant is given by the formula $a = \left[2 \times r\right]$. For example, the lattice constant of the SC-crystallized polonium is $\left[2 \times 0.167 \text{ nm}\right]$, or $0.334\, nm$. If the space lattice is FCC, the lattice constant is given by the formula $[4 \times r \ / \ (2)^{1/2}]$ and if the space lattice is BCC, then the lattice constant is given by the formula $a = [4 \times r \ / \ (3)^{1/2}]$.

References

- Kendrew, J. C.; Bodo, G.; Dintzis, H. M.; Parrish, R. G.; Wyckoff, H.; Phillips, D. C. (1958). "A Three-Dimensional Model of the Myoglobin Molecule Obtained by X-Ray Analysis". Nature. 181(4610): 662–6. Bibcode:-1958Natur.181..662K. doi:10.1038/181662a0. PMID 13517261.
- Triclinic-crystal-system-and-you: howlatm.com, Retrieved 25 March 2018
- Monoclinic-system, science: britannica.com, Retrieved 18 April 2018
- Annite-properties-and-occurrence, chemistry, science: assignmentpoint.com, Retrieved 24 March 2018
- Prince, E. (2006). International Tables for Crystallography Vol. C: Mathematical, Physical and Chemical Tables. Wiley. ISBN 978-1-4020-4969-9.
- Orthorhombic-crystal-system, crystal-shapes, earth-sciences-museum: uwaterloo.ca, Retrieved 11 May 2018
- Epsomite-properties-occurrences, geographic-minerals: assignmentpoint.com, Retrieved 11 June 2018
- Snigirev, A. (2007). "Two-step hard X-ray focusing combining Fresnel zone plate and single-bounce ellipsoidal capillary". Journal of Synchrotron Radiation. 14 (Pt 4): 326–330. doi:10.1107/S0909049507025174. PMID 17587657.
- Tetragonal-crystal-system-and-you: howlatm.com, Retrieved 31 March 2018

Permissions

All chapters in this book are published with permission under the Creative Commons Attribution Share Alike License or equivalent. Every chapter published in this book has been scrutinized by our experts. Their significance has been extensively debated. The topics covered herein carry significant information for a comprehensive understanding. They may even be implemented as practical applications or may be referred to as a beginning point for further studies.

We would like to thank the editorial team for lending their expertise to make the book truly unique. They have played a crucial role in the development of this book. Without their invaluable contributions this book wouldn't have been possible. They have made vital efforts to compile up to date information on the varied aspects of this subject to make this book a valuable addition to the collection of many professionals and students.

This book was conceptualized with the vision of imparting up-to-date and integrated information in this field. To ensure the same, a matchless editorial board was set up. Every individual on the board went through rigorous rounds of assessment to prove their worth. After which they invested a large part of their time researching and compiling the most relevant data for our readers.

The editorial board has been involved in producing this book since its inception. They have spent rigorous hours researching and exploring the diverse topics which have resulted in the successful publishing of this book. They have passed on their knowledge of decades through this book. To expedite this challenging task, the publisher supported the team at every step. A small team of assistant editors was also appointed to further simplify the editing procedure and attain best results for the readers.

Apart from the editorial board, the designing team has also invested a significant amount of their time in understanding the subject and creating the most relevant covers. They scrutinized every image to scout for the most suitable representation of the subject and create an appropriate cover for the book.

The publishing team has been an ardent support to the editorial, designing and production team. Their endless efforts to recruit the best for this project, has resulted in the accomplishment of this book. They are a veteran in the field of academics and their pool of knowledge is as vast as their experience in printing. Their expertise and guidance has proved useful at every step. Their uncompromising quality standards have made this book an exceptional effort. Their encouragement from time to time has been an inspiration for everyone.

The publisher and the editorial board hope that this book will prove to be a valuable piece of knowledge for students, practitioners and scholars across the globe.

Index